à
e

1.
tl
tc
re
it

24

25.

LF 9ڊ

Computer-Enhanced Analytical Spectroscopy

Volume 3

MODERN ANALYTICAL CHEMISTRY

Series Editor: David M. Hercules
University of Pittsburgh

A Continuation Order Plan is available for this series. A continuation order will bring delivery of each new volume immediately upon publication. Volumes are billed only upon actual shipment. For further information please contact the publisher.

Computer-Enhanced Analytical Spectroscopy

Volume 3

Edited by
Peter C. Jurs

The Pennsylvania State University
University Park, Pennsylvania

Plenum Press • New York and London

Library of Congress Cataloging in Publication Data

(Revised for vol. 3)

Computer-enhanced analytical spectroscopy.

  (Modern analytical chemistry)
  Papers from the First Hidden Peak Symposium, held at Snowbird, Utah, June 1986.
  Vol. 2 edited by Henk L. C. Meuzelaar.
  Vol. 2: Papers from the Second Hidden Peak Symposium on Computer-Enhanced
Analytical Spectroscopy held in June 1988 at Snowbird Resort, Utah.
  Vol. 3 edited by Peter C. Jurs.
  Includes bibliographical references and index.
  1. Spectrum analysis—Data processing. I. Meuzelaar, Henk L. C. II. Isenhour,
Thomas L. III. Jurs, Peter C. IV. Hidden Peak Symposium (1st: 1986: Snowbird, Utah)
V. Hidden Peak Symposium on Computer-Enhanced Analytical Spectroscopy (2nd:
1988; Snowbird, Utah). VI. Series.
QD95.C6323   1987   543′.0858   87-15883
  ISBN 0-306-43859-3

ISBN 0-306-43859-3

© 1992 Plenum Press, New York
A Division of Plenum Publishing Corporation
233 Spring Street, New York, N.Y. 10013

Printed in the United States of America

Contributors

Robert B. Bilhorn, Analytical Technology Division, Eastman Kodak Company, Rochester, New York

Dennis M. Davis, Analytical Division, Research Directorate, U.S. Army Chemical Research, Development, and Engineering Center, Aberdeen Proving Ground, Maryland

M. B. Denton, Department of Chemistry, University of Arizona, Tucson, Arizona

Stephen A. Dyer, Department of Electrical and Computer Engineering, Kansas State University, Manhattan, Kansas

Chris G. Enke, Department of Chemistry, Michigan State University, East Lansing, Michigan

William G. Fateley, Department of Chemistry, Kansas State University, Manhattan, Kansas

David M. Haaland, Sandia National Laboratories, Albuquerque, New Mexico

Robert M. Hammaker, Department of Chemistry, Kansas State University, Manhattan, Kansas

Peter Harrington, Department of Chemistry, Ohio University, Athens, Ohio

Kevin J. Hart, Oak Ridge National Laboratory, Oak Ridge, Tennessee

Huaying Huai, Center for Micro-Analysis and Reaction Chemistry, EMRC, University of Utah, Salt Lake City, Utah

Tomas Isaksson, Norwegian Food Research Institute, Ås, Norway

Peter C. Jurs, Department of Chemistry, Pennsylvania State University, University Park, Pennsylvania

Robert T. Kroutil, Analytical Division, Research Directorate, U.S. Army Chemical Research, Development, and Engineering Center, Aberdeen Proving Ground, Maryland

Steven P. Levine, Department of Environmental and Industrial Health, School of Public Health, University of Michigan, Ann Arbor, Michigan

Henk L. C. Meuzelaar, Center for Micro-Analysis and Reaction Chemistry, EMRC, University of Utah, Salt Lake City, Utah

Morton E. Munk, Department of Chemistry, Arizona State University, Tempe, Arizona

Tormod Næs, Norwegian Food Research Institute, Ås, Norway

R. S. Pomeroy, Department of Chemistry, University of Arizona, Tucson, Arizona

Miltiades Statheropoulos, Center for Micro-Analysis and Reaction Chemistry, EMRC, University of Utah, Salt Lake City, Utah

Sterling A. Tomellini, Department of Chemistry, University of New Hampshire, Durham, New Hampshire

Willem Windig, Chemometrics Laboratory, Eastman Kodak Company Rochester, New York

Barry J. Wythoff, Department of Chemistry, University of New Hampshire, Durham, New Hampshire

Yongseung Yun, Center for Micro-Analysis and Reaction Chemistry, EMRC, University of Utah, Salt Lake City, Utah

Preface

The Third Symposium on Computer-Enhanced Analytical Spectroscopy was held June 6–8, 1990 at the Snowbird Resort and Conference Center in the Wasatch Mountains outside of Salt Lake City, Utah. The program centered about two keynote lectures per session delivered by authorities in their fields. These keynote lectures have been gathered together here as chapters in the third volume of *Computer-Enhanced Analytical Spectroscopy*. The objective of this volume is to continue to provide a cross section of current research activities in this important and active field.

The overall purpose of this third CEAS symposium was to bring together experts on analytical spectroscopy and chemometrics to provide a status report on current research, to exchange information, and to stimulate discussion and cooperation. The lectures covered the application of computer methodologies to the practice of infrared spectroscopy, Hadamard spectroscopy, and near-infrared spectroscopy. Applications in mass spectroscopy and NMR spectroscopy were also covered. Talks on ion-mobility spectroscopy and atomic emission spectroscopy were also given. In addition to the formal lectures, posters were presented on related topics. These provided a focus for informal discussions during the meeting.

The third CEAS symposium was supported by a number of sponsors. Grateful thanks are given to the U.S. Army Chemical Research, Development, and Engineering Center, to the Analytical and Surface Chemistry Program of the National Science Foundation, to the Computers in Chemistry Division of the American Chemical Society, to the U.S. Environmental Protection Agency, to Merck Sharp & Dohme Research Laboratories, to the Department of Chemistry of Pennsylvania State University, and to the Center for Micro-Analysis and Reaction Chemistry of the University of Utah. Their support made the symposium possible.

The organizing committee for the third CEAS symposium included Charles L. Wilkins of the University of California, Riverside, and Joel Harris of the University of Utah. Special thanks go to Melinda Van for her help in running the symposium.

Peter C. Jurs

State College, Pennsylvania

Contents

Chapter 1

Multivariate Calibration Methods Applied to the Quantitative Analysis of Infrared Spectra

David M. Haaland

Chapter 2

Hadamard Methods in Signal Recovery

Stephen A. Dyer, Robert M. Hammaker, and William G. Fateley

Chapter 3

Computer-Assisted Methods in Near-Infrared Spectroscopy

Tormod Næs and Tomas Isaksson

Chapter 4

A Simple-to-Use Method for Interactive Self-Modeling Mixture Analysis

Willem Windig

Chapter 5

The Role of NMR Spectra in Computer-Enhanced Structure Elucidation

Morton E. Munk

Chapter 6

Computer-Assisted Mass Spectral Interpretation: MS/MS Analysis

Kevin J. Hart and Chris G. Enke

Chapter 7

Canonical Correlation Analysis of Multisource Fossil Fuel Data

Henk L. C. Meuzelaar, Miltiades Statheropoulos, Huaying Huai, and Yongseung Yun

Chapter 8

Developing Knowledge-Based Systems for Interpreting Infrared Spectra

Sterling A. Tomellini, Barry J. Wythoff, and Steven P. Levine

Chapter 9

Fuzzy Rule-Building Expert Systems

Peter B. Harrington

Chapter 10

Advanced Signal Processing and Data Analysis Techniques for Ion Mobility Spectrometry

Dennis M. Davis and Robert T. Kroutil

Chapter 11

Computerized Multichannel Atomic Emission Spectroscopy

Robert B. Bilhorn, R. S. Pomeroy, and M. B. Denton

Multivariate Calibration Methods Applied to the Quantitative Analysis of Infrared Spectra

David M. Haaland

1.1. Introduction

Multivariate calibration methods are beginning to have a major impact on the quantitative analysis of infrared spectral data. They have been shown to improve analysis precision, accuracy, reliability, and applicability of infrared spectral analyses relative to the more conventional univariate methods of data analysis. Rather than attempting to find and use only an isolated spectral feature in the analysis of spectral data, multivariate methods derive their power from the simultaneous use of multiple intensities (i.e., multiple variables) in each spectrum. Thus, the problem of spectral interferences can be eliminated with the use of any one of a number of multivariate calibration methods. A number of reviews of the various multivariate methods applied to infrared spectroscopy have recently appeared in the literature.[1–3] These methods include classical least squares[4] (CLS, also known as the **K**-matrix method), inverse least squares[5] (ILS, also known as the **P**-matrix method or sometimes called multiple linear regression, MLR), the **Q**-matrix method,[6] cross correlation,[7] Kalman

David M. Haaland • Sandia National Laboratories, Albuquerque, New Mexico 87185.

Computer-Enhanced Analytical Spectroscopy, Volume 3, edited by Peter C. Jurs. Plenum Press, New York, 1992.

filtering,[8] partial least squares[9-13] (PLS), and principal component regression[9, 10, 14, 15] (PCR). Because the last two methods often exhibit the greatest range of applicability and since they are beginning to find the largest following in the infrared community, more space in this chapter will be devoted to developing a qualitative understanding of these two factor analysis based methods.

A comparison of the four most heavily used multivariate calibration methods in infrared spectroscopy (i.e., CLS, ILS, PLS, and PCR) will be made using the Monte Carlo simulation results presented recently by Thomas and Haaland.[16] These simulations yield a basic understanding of the relative merits, power, and limitations of the four calibration methods under the condition that Beer's law is followed.

Finally, the importance of outlier detection will be discussed. Outliers are those samples among the calibration or unknown samples that are in some manner different from the rest of the calibration samples. Their inclusion in the calibration can lower the precision and accuracy of the analyses of all future unknown samples. The analysis of an outlier unknown sample may not be valid. Thus, it is crucial to the success of these multivariate methods that outliers be detected for both the calibration and unknown samples. The detection of outliers may also allow us to determine when a spectrometer has drifted out of calibration.

1.2. MODELS FOR VARIOUS MULTIVARIATE CALIBRATION METHODS

The differences between CLS, ILS, PLS, and PCR have been presented several times. These differences are a result of the formulation of the model relating spectral intensities and concentrations and the assumptions made in the models about the errors. In what follows, we use the convention that matrices are represented by uppercase bold letters, vectors are represented by lowercase italics letters. Vectors are expressed as column vectors, with row vectors being presented as transposed row vectors. Primes are used to represent transposed matrices and vectors.

1.2.1. Classical and Inverse Least-Squares Models

If the spectra are presented as rows of a matrix, then the CLS model which is based on the Beer–Lambert relation can be written as

$$\mathbf{A} = \mathbf{CK} + \mathbf{E_A} \tag{1.1}$$

where $\mathbf{A}$ is the $m \times n$ matrix of the absorbances of each of the m samples at n frequencies, $\mathbf{C}$ is the $m \times l$ matrix of the l component concentrations in the m samples, $\mathbf{K}$ represents the $l \times n$ matrix of pure-component spectra at unit concentration, and $\mathbf{E_A}$ represents the $m \times n$ matrix of spectral errors in the model (either noise or model error). The solutions to the calibration and prediction phases of the analysis are presented elsewhere.[3] In contrast, the ILS method is written as

$$\mathbf{C} = \mathbf{AP} + \mathbf{E_c} \qquad (1.2)$$

where $\mathbf{P}$ is the $n \times l$ matrix of regression coefficients relating the l component concentrations to the matrix of spectral intensities, and $\mathbf{E_c}$ is the $m \times l$ matrix of concentration errors in the model. If the concentration errors in the rows of $\mathbf{C}$ are uncorrelated, then Eq. (1.2) can be analyzed one component at a time, i.e.,

$$\mathbf{c} = \mathbf{Ap} + \mathbf{e_c} \qquad (1.3)$$

where $\mathbf{c}$ now represents the concentrations of a single analyte in the set of calibration chemical samples and $\mathbf{p}$ is the column of $\mathbf{P}$ that corresponds to the analyte of interest. Again, the solutions to Eqs. (1.2) and (1.3) have been presented elsewhere.[3]

It can be seen from Eqs. (1.1)–(1.3) that the assumptions and treatment of errors in CLS and ILS models are different. In CLS [Eq. (1.1)], the errors are assumed to reside in the spectral measurements, while in ILS [Eqs. (1.2) and (1.3)] the errors are assumed to arise from the concentration measurements only. Another consequence of the different forms of the three equations is that CLS requires a knowledge of all interfering chemical species while the ILS solution can be obtained one chemical component at a time if concentration errors are uncorrelated. In addition, the ILS solution can more readily be used for the estimation of chemical and physical properties of samples from their spectra. However, CLS can use all spectral intensities with potential improvements in precision while ILS is limited in the numbers of spectral intensities that can be included in the analysis when the least-squares solution of the ILS model is calculated.[17] ILS can accommodate all spectral intensities if data compression methods are used in solving the model.

1.2.2. Models for Factor Analysis Methods

Factor analysis methods such as PLS and PCR are data compression techniques that have separate steps to minimize spectral and concentration errors during the modeling of the calibration data. Both PLS and PCR use models which are of the form

$$A = TB + E_A \qquad (1.4)$$

$$c = Tv + e_c \qquad (1.5)$$

where the spectral matrix A has been factored into the product of two smaller matrices T and B with dimensions $m \times r$ and $r \times n$, respectively. The dimension of r is the rank of the matrix A which leads to the optimal model for concentration prediction. The rank r can be determined empirically from the calibration data.[9] The algorithms by which PLS and PCR perform the factor analysis are different and their handling of the error terms are also different. From Eq. (1.5), it can be seen that an inverse model similar to Eq. (1.3) has been incorporated into the PLS and PCR models. The difference between Eqs. (1.3) and (1.5) is that PLS and PCR have compressed the spectral matrix A into a new full-spectrum representation of the original data which can be described by a small number of intensities, T, in the full-spectrum basis vectors contained in the matrix B.

1.3. QUALITATIVE DESCRIPTION OF PLS AND PCR AND OUTLIER DETECTION METHODS

Since multivariate calibration methods based on factor analysis are the more versatile of the multivariate calibration methods available, it is important to obtain a qualitative understanding of how they work. In the following description, we discuss the concepts behind the calibration methods rather than providing the mathematical details of the algorithms that have been published elsewhere.[9, 15, 18] In this manner, a qualitative understanding of the methods, their power, and the diagnostic information derived from them may be more easily visualized. In Fig. 1.1, we present an example from a real data set which simply demonstrates the concepts involved in multivariate calibration methods based on factor analysis. The figure has been constructed using results of the partial least-squares (PLS) algorithm applied to the experimental data. However, the ideas presented are appropriate for any factor analysis calibration based on mathematically linear models. The primary factor analysis methods applied to infrared spectra include PLS and principal component regression (PCR).

1.3.1. Calibration

The data presented in Fig. 1.1 are from the analysis of phosphosilicate glass (PSG) thin films on silicon wafers. These films are used as dielectric

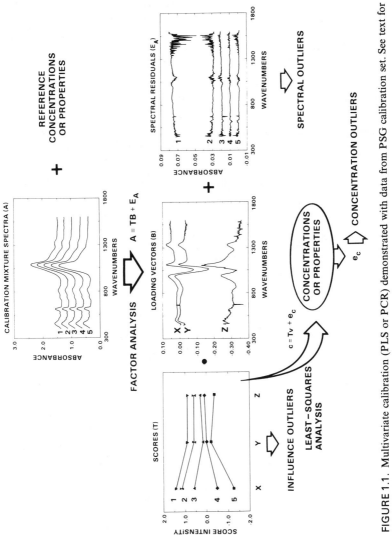

FIGURE 1.1. Multivariate calibration (PLS or PCR) demonstrated with data from PSG calibration set. See text for full details.

layers between conducting layers in the manufacture of integrated circuits. For quality control purposes, it is desired to control the phosphorus content of the thin-film glass in order to control the properties of the glass. The calibration information used in the analysis is presented at the top of the figure. The design of the calibration set is of extreme importance. In this simple case of a two-component system, the design is relatively straightforward with the phosphorus content increasing in a systematic manner with only minor variations in thickness of the films. If thickness variations were expected in future samples to be analyzed by these methods, then thickness should be varied in a systematic manner orthogonal to the phosphorus variation. These statistical designs have been discussed elsewhere.[19]

Displayed in the top box of Fig. 1.1 are five of 11 calibration sample spectra that were used in developing the partial least-squares (PLS) multivariate calibration model. The concentrations of phosphorus in the calibration samples were provided from an independent reference analysis method, an electron microprobe in this case. Chemical or physical properties of the samples could also be calibrated if these data were available for the calibration samples. The first step in the mathematical calibration of the data is the factor analysis of the spectral data. The factor analysis step decomposes the spectral matrix $\mathbf{A}$ into the product of two smaller matrices. These matrices, $\mathbf{T}$ and $\mathbf{B}$ [see Eq. (1.4)], consist of the scores and loading vectors, respectively, as described in factor analysis literature. Principal component regression (PCR) only uses the information in the spectral data to perform the factor analysis. In the first step of PCR, the principal component analysis of the spectral data generates loading vectors that, for a given number of loading vectors, will account for the maximum amount of the variance in the spectral data. PLS, on the other hand, incorporates concentration information into the factor analysis process and finds those loading vectors that maximize the covariance between concentration and spectra. Thus, PLS attempts to find loading vectors that correlate with concentration.

The loading vectors (comprising the rows of $\mathbf{B}$) in Fig. 1.1 result from a change in the coordinate system of the spectral data. We have transformed the spectral data matrix $(\mathbf{A})$ representing spectral intensities at more than 600 data points into a new full-spectrum representation of the spectral data requiring only three full-spectrum basis vectors (i.e., the three X, Y, and Z loading vectors in Fig. 1.1 which are the rows of $\mathbf{B}$). The selection of the number of loading vectors is empirically determined and is described elsewhere.[9] However, the number selected is dependent on the number required for optimal concentration predictions rather than the number required for optimal fitting of the spectral data. These vectors form a set of full-spectrum basis vectors which will reproduce each of the calibration spectra to within noise and model error (represented by $\mathbf{E}_A$ in the factor analysis step). The amounts of each of these loading vectors required to model each calibration spectrum represent the elements of the **T** matrix and are known as scores. Therefore, **T** is a new, highly condensed representation of the calibration spectral data. Thus, we have condensed over 3000 spectral intensities in Fig. 1.1 to a new representation of the data that contains only 15 intensities or scores. Of course to reproduce the model of the spectra and to analyze an unknown spectrum, the three loading vectors must also be retained.

The scores now can be considered spectral intensities in the new full-spectrum coordinate system of the loading vectors. These scores retain all the information that is important for phosphorus concentration prediction contained in the original calibration spectra (assuming the linear model holds). The scores associated with each sample have been shifted in Fig. 1.1 for clarity. Also, the scores for each sample have been connected by lines in Fig. 1.1 to make the data appear like spectra. This presentation of the score data emphasizes that the scores can be considered new spectral representations of the original spectra. Because each of the three scores for each calibration sample contains information accumulated from all 600 intensities of the original calibration spectrum, there is a significant signal-to-noise enhancement for these individual scores relative to the individual spectral intensities of the original spectral data. The actual amount of improvement in signal-to-noise is dependent on both the numbers of data points and their relative intensities. With these improvements, it is clear that these multivariate methods can enhance the analysis precision relative to frequency-limited analyses by signal averaging over a wide range of spectral intensities. With this improvement in precision using full-spectrum methods, we have been able to quantitate concentrations from data that have signal-to-noise ratios less than one at any frequency in the spectrum.[20]

Once the scores are obtained, they can be related to analyte concentrations using the inverse least-squares model presented in Fig. 1.1. The elements of the vector **v** simply represent the proportionality constants (or regression coefficients) which linearly relate the scores to concentrations. This least-squares step completes the multivariate calibration of the data.

1.3.2. Outlier Detection during Calibration

With the calibration complete, we can now use powerful statistical methods to evaluate the quality of the model and to identify outlier or problem samples. The multivariate nature of the data provides us with significant information for performing several diagnostic procedures. If a method called cross validation is used when building the calibration model,[9, 21] then the precision expected for the analysis of future unknown samples can be determined. In addition, outliers present in the calibration set can be detected.[9–11, 22] Calibration outliers are those spectra or data in the calibration which are, in some manner, different than the rest of the calibration data. Outlier samples in the calibration set may not follow the model developed from the remaining calibration samples. Therefore, their presence in the calibration may contaminate the calibration and decrease precision and accuracy. It is important to detect outliers in the calibration set, since their presence may reduce the precision and accuracy of the analyte concentration determination in all future unknown samples. Their presence also reduces sensitivity to outliers in the analyses of unknown samples. Figure 1.1 illustrates where in the multivariate calibration process that outlier detection can be accomplished.

1.3.2.1. Spectral Residuals

The spectral residuals in matrix $\mathbf{E_A}$ offer one of the most sensitive measures for outlier detection. The spectral residuals represent the difference between the measured spectra and the spectra obtained from the calibration model. The spectral residuals displayed in Fig. 1.1 are not actually from the full calibration model, but rather from the cross-validated procedure whereby each calibration sample has been separately removed from the calibration and predicted based on the calibration of the remaining calibration samples. Therefore, each spectral residual is actually the spectral residual obtained if this sample were an independent unknown sample. This cross-validation procedure offers greatly increased sensitivity to outliers as will be discussed later. A spectroscopist can often interpret these spectral residuals and identify why a particular sample is an outlier. For example, the presence of an unexpected component in the sample can often be identified since the spectrum of the unexpected component may be reflected in the spectral residual.

The spectral residuals presented in Fig. 1.1 are scale expanded by a factor of 30 over the scale of the original calibration spectra in order to observe structure in the spectral residuals. The major features in the spectral residuals are due to water vapor and CO_2 present in the purge gas of the spectrometer. Also present is evidence for a slight lack of model fit for the strong $Si-O$ stretching vibration which may be due to greater spectrometer nonlinearities experienced by strongly absorbing bands. The presence of the purge species in the spectral residuals might be expected since the spectra of these gases are present in varying amounts, but they do not correlate with the phosphorus concentrations of the samples. Clearly, we can interpret the spectral residuals in this calibration.

1.3.2.2. Spectral F Ratios

For automatic outlier detection, we must currently rely on a measure of the overall magnitude of the spectral residuals to identify specific outlier samples. An automated outlier detection method we use involves the calculation of spectral F ratios based on the sum of squared residual intensities for each spectrum relative to the averaged sum of squared residual intensities for the remaining calibration spectra.[9] The F ratio, therefore, is a statistical measure of the size of the spectral residuals relative to the average spectral residual for the calibration set. If errors are uncorrelated and normally distributed, then the value of the F ratio can be used to determine a probability that the spectrum follows the model of the remaining calibration samples. In practice, the spectral residuals are often correlated, and we find empirically that F ratios above three are suggestive of outlier samples when the number of frequencies included in the analysis is large and spectral errors are correlated.[9] Using the F ratio, we can then have the multivariate calibration software automatically flag spectral outliers.

In the example depicted in Fig. 1.1, the spectral residual for sample 1 has been flagged as an outlier by the F statistic during cross validation (the F ratio is 5.3). However, the spectral residuals in this case are large only as a result of the large amount of water vapor and CO_2 present in the spectrum. Since these gases are present in the purge gas rather than the sample, and because these features are not related to phosphorus content, it might be expected that this is not an important outlier to delete from the calibration. In fact, analysis precision was not improved by the deletion of this sample from the calibration. Therefore, this outlier sample was retained in the calibration set.

1.3.2.3. Mahalanobis Distance

Another useful outlier detection method is based on the scores for each sample. One determines a distance of each sample in score space from the center of the calibration score data. If a sample is located at a distance that is far removed from the center of the score space for the bulk of the calibration samples, then it is a relatively unique sample which may not follow the model based on the remainder of the calibration samples. That is, its scores are outside the range of the other calibration samples, and extrapolation of the model would be required. Extrapolation is never expected to be as reliable as interpolation. The Mahalanobis distance is a useful measure for the distance of the scores from the center of the calibration.[15] The Mahalanobis distance is different from the Euclidian distance in that each dimension in the distance is scaled to the standard deviation of the data in that dimension. It is the appropriate distance to use if the calibration concentration errors are multivariate normally distributed and if the concentration errors are large relative to the errors in the scores (i.e., concentration errors large relative to the spectral errors averaged for the analyte contribution over the spectral analysis region).

When using the full calibration model, the Mahalanobis distance gives a measure of the influence (sometimes called leverage) of each sample on the calibration model. If the sample is an extreme sample, it will have a large influence on the model and therefore may require a new factor to have its spectrum appropriately modeled. In this case, the Mahalanobis distance will be close to one. Samples very near the center of the scores will exert little influence on the model and will have a Mahalanobis distance near 0. Therefore, we call an outlier detected by this scheme an influence outlier. In addition, the Mahalanobis distance can be used to obtain estimates of the precision of the analysis for any sample which follows the calibration model.

1.3.2.4. Concentration F Ratio

Another source of outlier detection can be obtained from a comparison of the differences between the reference concentrations and the concentrations estimated from the calibration model (i.e., e_c in Fig. 1.1). If any of these differences are relatively large, then an outlier may be indicated. In practice, we form a F ratio which compares the leverage-corrected squared concentration residuals to the average leverage-corrected squared concentration residuals for the remainder of the calibration samples.[23] The residuals are leverage corrected to account for the fact that poorer predictions are expected in least-squares analyses for samples whose spectra are far from the average of the calibration data. In essence, this outlier detection method is asking which of the calibration predictions are larger than expected from the model given the presence of random errors in the data. Often, outlier samples detected by this outlier method are the result of errors in the analysis of the sample by the reference method. This is especially likely if the same sample is not flagged as an outlier by the other outlier detection methods. Re-analysis of the sample by the reference method would be appropriate in this latter case.

1.3.2.5. General Comments about Calibration Outlier Detection

In combination, these three outlier detection schemes are very useful in characterizing the quality of the calibration set. By eliminating known problem samples, future predictions on unknown samples can be made more precise and accurate, and outlier detection is rendered more sensitive.

However, care must be exercised in eliminating calibration samples. If a sample is an outlier because it is an extreme sample, then it must be decided if this extreme region of the calibration is important for future unknown samples. If it is, then rather than deleting the outlier sample, a better approach would be to add samples to the calibration set which will further expand, explore, and model this region of the calibration space.

1.3.3. Qualitative Interpretation of Loading Vectors

It is interesting to note that the loading vectors obtained during the calibration contain very useful qualitative information that spectroscopists can interpret. For PLS, the first weight loading vector is the best least-squares approximation to the pure phosphorus oxide spectrum as it exists in the silica matrix over the concentration range of the calibration.[9] The first PLS loading vector is derived from the first weight loading vector and, in the case presented in Fig. 1.1, the two are nearly identical. Thus, the broad positive peaks in the first PLS loading vector represent the regions of the spectrum that are positively correlated with the phosphorus component. The single negative peak is the result of phosphorus displacing the silica component as the phosphorus concentration increases. The original PCR loading vectors are less likely to be as interpretable as the PLS loading vectors, since they are obtained without any direct information about the phosphorus content of the samples. The second loading vector in the figure is quite similar to the first except for a shift in the vector position and a slight slope change. This vector is primarily used by the model to correct for linear base-line variations between the calibration spectra. In combination, the loading vectors labeled X and Y in Fig. 1.1 can independently account for phosphorus and base-line changes in the calibration spectra. The third factor (Z in Fig. 1.1 which has been shifted in the figure relative to X and Y for clarity) appears to be required to account for both deviations in Beer's law and additional base-line slope variations. Thus, the relative intensities of the two phosphorus bands on the side of the $Si - O$ stretching vibration do not vary in the same proportion as changes occur in phosphorus concentration. This is a clear indication that Beer's law is not followed for this system.

1.3.4. Prediction

In Fig. 1.2, we present a similar pictorial representation of the multi-variate prediction of an unknown sample based on the calibration model developed during the calibration phase of the analysis. Once the spectrum of the unknown sample is obtained, the calibration model can be applied to the spectrum. Using the X, Y, and Z loading vectors from Fig. 1.1, a

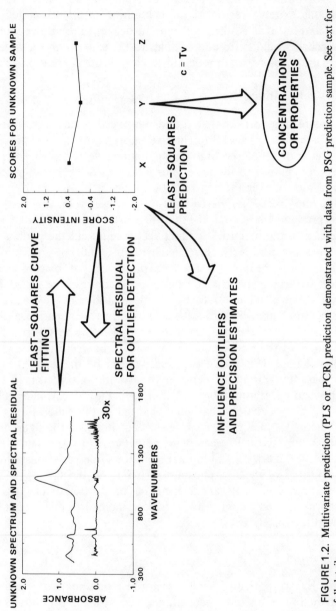

FIGURE 1.2. Multivariate prediction (PLS or PCR) prediction demonstrated with data from PSG prediction sample. See text for full details.

least-squares curve-fitting procedure is performed to obtain a fit of the loading vectors to the unknown sample spectrum. The amounts of each of the loading vectors used to give the least-squares fit of the spectrum are the scores, i.e., the intensities of the unknown spectrum in the new full-spectrum coordinate system of the loading vectors. These scores are depicted on the right side of Fig. 1.2. Given the scores, the concentration of the analyte may be estimated directly using the v vector that was determined during the inverse least-squares calibration.

1.3.5. Outlier Detection during Prediction

The difference between the measured and modeled spectrum is shown on the left side of Fig. 1.2 under the unknown sample spectrum. These spectral residuals have been scale expanded by a factor of 30 relative to the scale of the unknown sample spectrum. This residual spectrum can again be interpreted by the spectroscopist and can be used to calculate a F ratio for the unknown sample. The F ratio value of 0.5 for this unknown sample does not indicate that this particular sample in Fig. 1.2 is a spectral outlier. The scores can be used to determine if the sample is an extreme sample relative to the calibration samples. If the sample is not flagged as a spectral outlier by the spectral F test, then the sample is expected to follow the calibration model, and an estimate of the precision of the concentration estimate may be made from the sample's Mahalanobis distance.[23] This estimate of precision becomes larger as the sample's scores extend further from the center of the scores of the calibration samples. When using PCR, this precision estimate should be reasonable for samples following the calibration model if the data are multivariate normally distributed and if the spectral errors in the calibration and unknown spectra are small relative to the errors in the concentrations determined by the reference method used in the calibration. In the case of PLS, the quality of the precision estimates may be degraded by the influence of concentration errors on the scores. Nevertheless, the precision estimates can be used as guidelines for PLS analyses also.

1.3.6. Results for PSG Example

The results of a cross-validated PLS calibration of phosphorus weight percent are compared in Fig. 1.3 with the reference concentrations obtained by electron microprobe analysis for the 11 calibration samples used in obtaining the results in Fig. 1.1. The standard error of prediction (SEP) based on the cross-validation calibration leaving one sample out at a time is 0.09 wt% phosphorus for this set of samples. This is comparable to the estimated precision of the reference method of 0.1 wt%. The PLS analysis

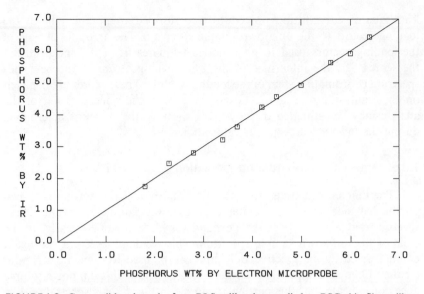

FIGURE 1.3. Cross-validated results from PLS calibration applied to PSG thin-film calibration samples.

precision is therefore limited by the precision of the reference method. The precision demonstrated in Fig. 1.1 should be expected for future unknown PSG samples that are within the range of the calibration and are not determined to be outlier samples.

1.4. COMPARISON OF PREDICTION ABILITIES OF CLS, ILS, PLS, AND PCR

1.4.1. Monte Carlo Simulations

Given the existence of a variety of multivariate calibration methods, the spectroscopist might ask which method is the best to use or, if no single method is always best, which method should be used with a given set of spectral calibration data. Thomas and Haaland[16] have attempted to address this question with a series of Monte Carlo simulations that span a range of parameters which were thought to possibly affect the relative predictive performance of the various calibration methods. They used simulated data that followed the Beer's law model and obeyed most of the assumptions contained in the CLS, ILS, PLS, and PCR models. Using a set of simulations that corresponded to a set of three-component mixtures with the constraint that the sum of the mole fractions equals one, Thomas and Haaland studied the effect of eight different experimental factors which might affect the relative predictive performance of the four multivariate calibration methods. The simulations were performed using a two-level, eight-factor factorial design with sufficient numbers of calibrations and predictions to be able to determine differences between the methods with high statistical significance. The factors investigated and the values selected for the two levels are shown in Table 1.1. Spectra of the pure components and representative spectra for the cases where the relative spectral intensities of the pure-component spectra are 1.0:0.1:0.1 are shown in Fig. 1.4 and Fig. 1.5, respectively. For ILS where a limited number of frequencies must be selected, a limited all-sets search was made to find those frequencies which yielded the best prediction precision for ILS.

1.4.2. Prediction Performance Abilities

General conclusions showed that the three full-spectrum methods (CLS, PLS, and PCR) almost always outperformed the frequency-limited ILS method. This is because the full-spectrum methods could take advantage of the signal averaging effect obtained when multiple intensities with redundant information are included in the analysis. Only when concentration noise was present (i.e., 5% relative concentration errors in the reference method) and when the analyte was a major spectral component did ILS have a prediction precision comparable to the full-spectrum methods. The results of the main effects on the prediction abilities of the four methods are summarized in Figs. 1.6–1.8. The log of the mean squared prediction error [log(MSPE)] is presented at the left of Figs. 1.6–1.8 for the benchmark conditions where all factors are at their least degrading conditions. The MSPE is calculated from the sum of squared differences between the estimated and true concentrations. The cases with the smallest prediction errors are represented by the bars with the most negative

TABLE 1.1. Factor Levels

Factor	Label	Low level	High level
Concentration noise	(X_1)	None	$\sigma_c = 0.02$ mole fraction
Spectral noise	(X_2)	$\sigma_a = 0.001$ au	$\sigma_a = 0.005$ au
Separation of spectral features	(X_3)	25 cm^{-1}	12.5 cm^{-1}
Spectral base-line variation	(X_4)	None	Random linear
Number of intensities per spectrum	(X_5)	200	25
Calibration set configuration	(X_6)	Designed	Random
Number of calibration samples	(X_7)	15	7
Pure-component intensities for components A:B:C	(X_8)	1:0.1:0.1	1:1:1

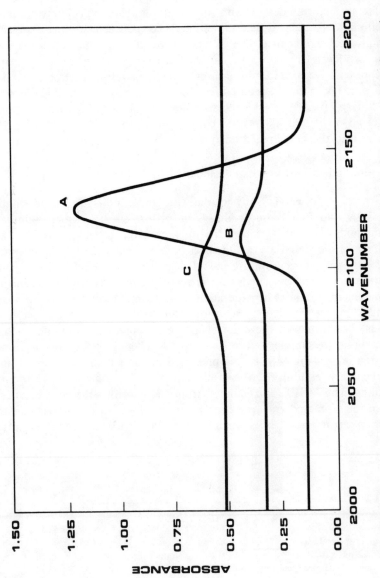

FIGURE 1.4. Pure-component spectra when intensities for components A:B:C are in the ratio 1:0.1:0.1 and band separation is at a minimum. (Reprinted with permission from Ref. 16. Copyright 1990 American Chemical Society.)

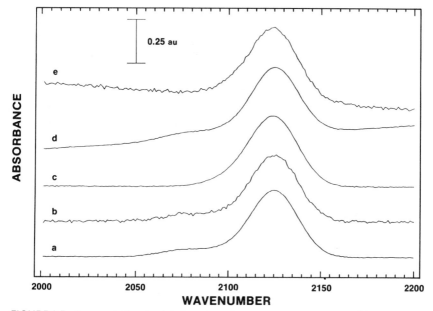

FIGURE 1.5. Representative simulated spectra when pure component spectral intensities are 1:0.1:0.1 and concentrations are 1/3:1/3:1/3. (a) Spectral noise, spectral overlap, and random linear base-line factors are at their least degrading conditions. (b) Spectral noise is at the high level. (c) Spectral overlap is at its more degrading condition. (d) Linear base line has been added. (e) Spectral noise, spectral overlap, and random base-line factors are all at their more degrading condition. (Reprinted with permission from Ref. 16. Copyright 1990 American Chemical Society.)

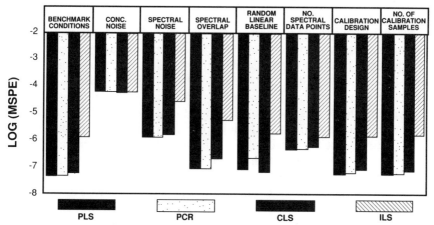

FIGURE 1.6. Log(MSPE) for each of the multivariate calibration methods for the major spectral component (A) when all factors are at their benchmark (least degrading) conditions and when each factor is separately set to its more degrading condition. (Reprinted with permission from Ref. 16. Copyright 1990 American Chemical Society.)

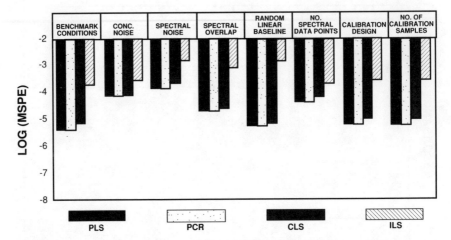

FIGURE 1.7. Log(MSPE) for each of the multivariate calibration methods for the minor spectral component (C) when all factors are at their benchmark (least degrading) conditions and when each factor is separately set to its more degrading condition. (Reprinted with permission from Ref. 16. Copyright 1990 American Chemical Society.)

log(MSPE). The log(MSPE) is also presented in the figures for the case where each of seven factors is separately switched to its more degrading level while all other levels are at their less degrading condition.

From Fig. 1.6, it is clear that for the range of simulations studied, concentration errors have the greatest degradation effect on the concentration predictions for the major spectral component. In this case, the relatively large concentration errors also have the ability to make the prediction performance of the major spectral component the same for all methods. Figure 1.7 shows that the effect of concentration noise is not as great for the minor spectral components. Thus, the signal averaging effect of the full-spectrum methods is sufficiently important for these components with low signal-to-noise ratios, that the concentration errors are not dominant for these minor spectral components. Nevertheless, the precision of the reference methods must be relatively high for us to take full advantage of the precision enhancement of the full-spectrum multivariate methods.

Spectral noise is the second most important factor affecting prediction performance in these simulations given the range of variation present in the eight factors. In addition, the order-of-magnitude decrease in spectral intensities for the minor spectral component causes a decrease in MSPE of approximately two orders of magnitude as observed when comparing Figs. 1.6 and 1.7. Therefore, signal-to-noise ratios in the spectra can be very important factors in the resulting analysis precision of these methods.

The effect of random spectral base lines that are at most 20% of the maximum spectral intensity is shown in Figs. 1.6 and 1.8 to cause a

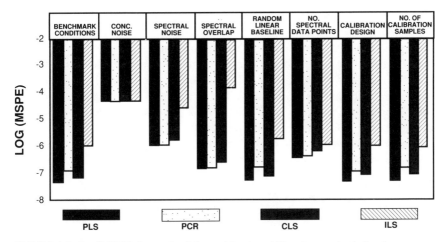

FIGURE 1.8. Log(MSPE) for each of the multivariate calibration methods for the component (A) when all pure-component spectra have equal intensities. Presented is the case when all factors are at their benchmark (least degrading) conditions and when each factor is separately set to its more degrading condition. (Reprinted with permission from Ref. 16. Copyright 1990 American Chemical Society.)

statistically significant difference between PLS and PCR predictive abilities. This difference in PLS and PCR might be expected, since the PLS loading vectors have been constructed to increase concentration predictive ability relative to PCR in the presence of spectral variations that are not correlated with concentrations of the analyte. CLS is not degraded in this case with random spectral base lines, since we simultaneously fit a linear spectral base line with the CLS algorithm. However, if random spectral base lines were present which did not follow the linear model, CLS would degrade in performance relative to the other three methods. The presence of the randomly varying third major spectral component is observed in Fig. 1.8 to have the same effect as the randomly varying spectral base line on the relative performance of PLS and PCR.

ILS is seen to approach the performance of full-spectrum methods as the number of spectral data points is decreased for the full-spectrum methods. This is reasonable, since the ILS method uses the same number of intensities in each case while the full-spectrum methods have a significant reduction in spectral averaging as the number of intensities approaches the minimum required for prediction. In the limit where the number of spectral intensities is less than or equal to the number required for optimal ILS predictions, all methods will have similar predictive ability.

Although not discernible from the figures, it has also been found from the simulations that there is an indication that PLS outperforms PCR for minor spectral components, and PCR outperforms PLS when concentration errors are high. The former might be expected, since PLS calculates those loading vectors that maximize covariance between concentrations and spectral data. Therefore, more spectral information from minor spectral features might be expected to occur in a given number of loading vectors for PLS relative to PCR. The latter is reasonable, since PLS uses concentration information both in the factor analysis of the spectra and the regression of concentrations on scores. PCR, on the other hand, only uses concentration information in the regression of concentrations on scores. The extra involvement of concentrations in the PLS analysis might be detrimental when concentration errors are relatively large.

1.4.3. Additional Considerations

Although the above simulation results yield considerable insight into the relative performance of the four multivariate calibration methods, they are strictly valid only for those cases where Beer's law is followed and for independent and normally distributed errors in spectra and concentrations. For real data, these assumptions may not be valid and alternate conclusions might be reached. In unpublished results,[24] we find that for a variety of types of deviations in Beer's law, PLS and PCR yield comparable predictive ability, and both outperform ILS and CLS. With some deviations in Beer's law, ILS actually outperforms CLS methods since CLS is more dependent on Beer's law. The presence of errors that are not normally distributed has not been seen to affect the general conclusions presented in this chapter. It should be stated that it is possible that for samples where only a small fraction of the spectra follow Beer's law, it is possible for ILS to outperform the full-spectrum methods. However, the full-spectrum methods yield additional information which make them superior to the frequency-limited ILS method. That is, the full-spectrum methods allow the spectroscopist to examine full-spectrum residuals which provide enormous diagnostic information for both the calibration and unknown spectra. They also yield significantly improved outlier sensitivity, since there are more spectral intensities used in the outlier detection procedure. The increased outlier detection capabilities and greater interpretable diagnostic information make the full-spectrum factor analysis methods of PLS and PCR much more valuable in the analyses of real data and spectra.

1.5. IMPORTANCE OF CROSS VALIDATION IN OUTLIER DETECTION

The detection of outliers is extremely important in the calibration phase of the multivariate analysis of spectral data. If outliers are not detected and handled in an appropriate manner, the analysis precision and accuracy of all future samples will be degraded. In addition, the presence of outliers in the calibration tends to inflate the base-line measures used in the detection of outliers, resulting in less sensitive tests for the detection of outliers among the unknowns. Unfortunately, the detection of outliers is also less sensitive during calibration if cross-validation procedures are not used. However, since cross validation is recommended for both selecting the optimal model as well as estimating the prediction ability of the factor analysis methods, cross-validation results should be available for outlier detection. Further demonstrations of the value of cross validation in outlier detection are made using another example of importance to the micro-electronics industry.

1.5.1. Borophosphosilicate Glass (BPSG) Example

The PLS and PCR analysis of borophosphosilicate glass (BPSG) thin films has been presented previously.[11, 25] These glasses are very important in the manufacture of integrated circuits, and monitoring their composition is important for quality control of the devices produced. The calibration set consisted of samples whose concentrations of boron and phosphorus ranged from 1 to 5 wt% and from 2 to 6 wt%, respectively. Film thickness ranged from 430 nm to 1000 nm. Figure 1.9 shows 60° incident angle spectra of four of the 44 calibration samples. These four spectra exhibit the extremes in

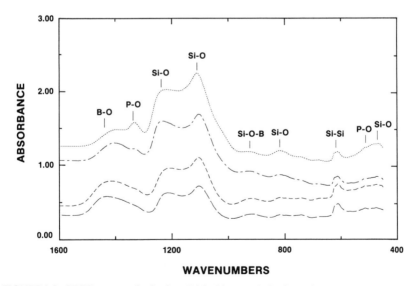

FIGURE 1.9. BPSG spectra obtained at 60° incident angle for four of the calibration samples exhibiting extremes in spectral differences.

spectral variations for the calibration set. This three-component system is useful in demonstrating the importance of cross-validation approaches to outlier detection. The cross-validated PLS analysis precision of boron in the 44 annealed BPSG thin-film samples was found to be 0.11 wt % (SEP). The prediction ability of the PLS calibration based upon the 1600–450 cm^{-1} region of the calibration BPSG spectra obtained at 60° incident angle is shown in Fig. 1.10.

1.5.2. Detection of Outliers in the BPSG Calibration

When outlier detection is based upon the full-calibration model, none of the BPSG calibration samples is detected as an outlier by spectral F ratios. However, when cross validation is used during calibration, then sample 6 is clearly detected as a spectral F-ratio outlier. Sample 36 is detected as a concentration F-ratio outlier independent of whether the outlier detection involves the cross-validated or full-calibration model. In this example, samples had been prepared in duplicate. Therefore, cross validation was performed by removing the duplicate pairs rather than single samples.

Spectral F-ratio results are demonstrated in Figs. 1.11 and 1.12 for the full-calibration and cross-validated results, respectively. Significance levels are difficult to attribute to spectral F ratios, since the presence of correlated

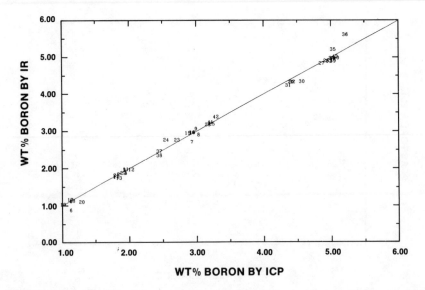

FIGURE 1.10. Cross-validated results from PLS calibration applied to BPSG thin-film calibration samples.

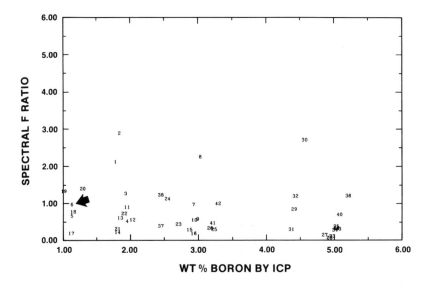

FIGURE 1.11. Spectral *F* ratios for full PLS calibration model for the BPSG calibration set. Arrow marks outlier sample 6.

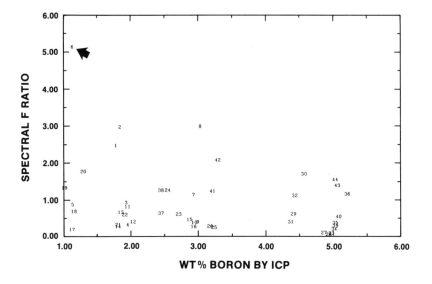

FIGURE 1.12. Spectral *F* ratios for cross-validated PLS calibration model for the BPSG calibration set. Arrow marks outlier sample 6.

spectral errors violates the assumptions made in converting spectral F ratios into probability levels. However, a spectral F ratio greater than 5 is clearly an outlier since the sum of squared spectral residuals for sample 6 is more than five times the average obtained from the cross-validated models based on the other 43 samples. The fact that the full calibration was unable to detect any outliers is attributed to the fact that sample 6 is a significant outlier which is quite different from the other calibration samples. Therefore, during the full calibration, sample 6 has such a large influence on the calibration model that it is actually predicted quite well when included in the calibration. However, when it is left out of the calibration, it has no influence on the calibration results, and it can then be detected as clearly different from the other calibration samples.

The cross-validated concentration F ratio for sample 34 is 13, which indicates significance at the 0.001 level. That is, there is only one chance in a thousand that a F ratio would be as large as 13 due to random noise for this sample set containing 44 samples.

1.5.3. Identification of Reasons for Samples Being Detected as Outliers

We would like to determine why samples 6 and 36 are outliers. An examination of cross-validated spectral residuals shows that the $Si-O$ stretching vibration of sample 6 is shifted to lower frequency relative to the

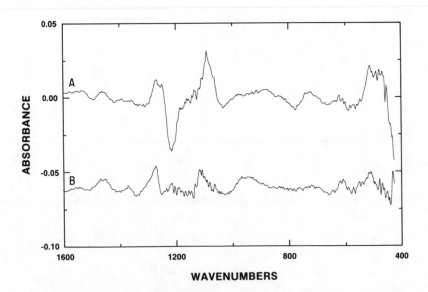

FIGURE 1.13. PLS spectral residuals based on cross-validated calibration models for BPSG samples 6 (labeled A) and 5 (labeled B) deposited under identical conditions.

model developed from the other 43 calibration samples. This might suggest that sample 6 had received a different annealing treatment during thermal processing than the other BPSG thin-film samples. The shift could also arise from a shift in the angle of incidence to less than 60° for this particular sample. Although the interpretation of the spectral residuals is not unique, the information in the spectral residuals narrows the options for exploring the cause of the sample spectrum being flagged as an outlier. The difference in the spectral residuals between the cross-validated and full-calibration models is shown in Figs. 1.13 and 1.14 for sample 6 and sample 5, which was deposited under the same conditions as sample 6 and therefore should have been a replicate for sample 6. Sample 5 was not flagged as an outlier. The full-calibration spectral residuals for samples 5 and 6 are of comparable intensity. Clearly, the influence of sample 6 in the full calibration makes it impossible to detect sample 6 as an outlier or to reliably interpret its spectral residual.

Sample 36 may be an outlier simply because of a poor analysis by the reference method. Unfortunately, the reference method was destructive, and the sample was no longer available for a repeat analysis. It may have also been an outlier because it is a relatively extreme sample. Sample 36 has the highest boron content and nearly the smallest thickness and lowest phosphorus content among the calibration samples. However, it is not an extreme sample in spectral space since the spectral F ratio is not large and its Mahalanobis distance based on the spectral scores is not extreme.

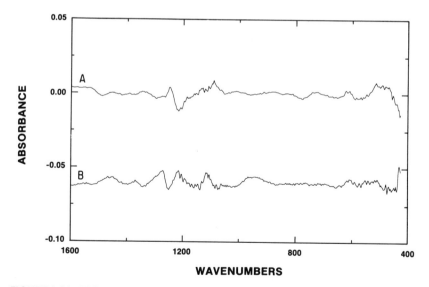

FIGURE 1.14. PLS spectral residuals based on the full-calibration model for BPSG samples 6 (labeled A) and 5 (labeled B) deposited under identical conditions.

1.5.4. Cautions about Outlier Detection

This example illustrates that while it is possible to detect outlier samples from the full-calibration model, a cross-validation detection of outliers is a more sensitive and reliable method for detecting outliers in the calibration set. It can similarly be shown that outlier detection and estimates of sample precision for the unknown samples are more realistic for a cross-validated model than those based on the full-calibration model. However, it should be emphasized that cross validation is not foolproof in the detection of outliers. Cross-validated detection of outliers is made less sensitive to outliers if multiple samples are outliers for similar reasons. Thus, when replicate samples or replicate spectra are present in the calibration, all replicates should be removed simultaneously during the cross-validation procedure. Otherwise, any replicates remaining in the calibration will be able to adequately model the removed outlier sample(s) with the risk that none of the replicates will be detected as outliers and the prediction ability of the multivariate calibration will be overly optimistic for the analysis of future unknown samples. This is the reason the duplicate pairs in the BPSG example were removed during the cross-validation process.

Since the detection of outliers is crucial to the development of precise, accurate, and reliable calibrations and predictions, all available outlier detection tools should be used. There is room for further improvement and development of new outlier detection methods.

1.6. SUMMARY

Multivariate calibration methods have greatly extended the capabilities of quantitative spectral analyses. No longer must the spectroscopist find an isolated spectral feature for each analyte in order to perform the analysis. By employing all the useful spectral information, the spectroscopist can achieve higher precision, greater accuracy, and increased reliability even in those cases where the spectrum of the analyte is overlapped with other components. Many of these multivariate methods can also be used to develop models for estimating chemical and physical properties of the samples.[14] In addition to the quantitative information, the full-spectrum multivariate calibration methods have been shown to contain useful qualitative information as well. Previously, it had been shown that CLS yields important qualitative information in the least-squares-estimated pure-component spectra.[4] In the first example presented in this chapter, PLS has also been shown to contain interpretable information in the weight loadings and loading vectors. All the full-spectrum methods can generate spectral residuals which can often be interpreted by the spectroscopist.

These spectral residuals are very useful for other diagnostic information, such as the detection of outlier samples among the calibration and unknown sample sets. The frequency-limited ILS method does not contain the same wealth of qualitative information or sensitivity to outliers that is possible with the full-spectrum methods.

The advantages in prediction ability of the full-spectrum methods were illustrated with Monte Carlo simulations comparing CLS, ILS, PLS, and PCR. These simulations were instructive for identifying those parameters that affect the relative performance of the various methods when the model assumptions are followed. The advantages of the PLS and PCR factor analysis methods are observed to be dramatic when all components are not known in the calibration set and when calibration concentration errors are not large. Although more can and will be learned about the relative performance of the methods with real data sets, the simulations provide a much larger range of variation in data sets than can studied in practice with real data.

The importance of outlier detection cannot be overemphasized. The presence of outliers in the calibration set reduces accuracy and precision and makes the detection of future outliers less sensitive. When these multivariate methods are performed in an automated mode, the analysis of unknown samples cannot be considered reliable if sensitive outlier detection is not present. It has also been shown that outlier detection can be improved with the use of cross-validated calibrations. Cross validation is important, since extreme outliers in the calibration can influence the full-calibration model so much that outliers become difficult to detect. Sensitive outlier detection should also identify when the spectrometer has drifted out of calibration.

An area of study that has just recently received attention is the issue of the transferability of the calibration model between different spectrometers. Since the calibration process is usually expensive and time consuming, transfer of the calibration model might be preferred over transfer of the calibration samples. This calibration transfer becomes crucial when the calibration samples are unstable with time. Sensitive outlier detection will most often allow the spectroscopist to identify if a calibration model is yielding reliable results on another spectrometer. However, since the full-spectrum calibration methods are usually sensitive to minor differences in spectra, spectrometer-dependent spectral differences will often cause analysis problems. A variety of methods can be proposed to handle this transferability problem. The most important might be to identify those spectrometer-dependent parameters which limit the transferability. These parameters can then be more carefully controlled between spectrometers. It is also possible to randomly split the calibration samples and obtain a portion of the spectra from the calibration set on each spectrometer in order to incorporate specific spectrometer differences in the calibration. Another method to improve transferability would be to simply adjust the calibration model based upon the spectra of a fewer number of calibration samples taken on the second spectrometer. Finally, a spectrometer transfer function might be determined from a subset of calibration samples in order to preprocess the spectral data to make it appear as if the spectra had been obtained on the original spectrometer. Clearly, transferability of calibrations is an area that will receive greater attention as multivariate calibration methods are used more heavily in industrial process monitoring and quality control settings.

Finally, it is clear that the use of multivariate spectral calibrations will increase in the future. Their use in quality control and process monitoring represents tremendous growth potential. Hopefully, the qualitative picture of the factor analysis multivariate methods presented in this chapter will increase the nonspecialist's knowledge and familiarity with these multivariate methods and will facilitate their future increased use and success.

ACKNOWLEDGMENTS

This work was performed in part at Sandia National Laboratories supported by the U.S. Department of Energy under Contract Number DE-AC04-76-DP00789 and by Semiconductor Research Corporation/ SEMATECH through the University of New Mexico Sematech Center of Excellence.

J. Linn of Harris Corp. is gratefully acknowledged for providing the PSG samples, their IR spectra, and the phosphorus reference concentrations. M. Kay provided the BPSG samples and C. E. Case aided in the collection of their IR spectra. E. V. Thomas aided in the multivariate software development, implementation of outlier detection methods, and provided statistical consultations. D. K. Melgaard made significant improvements in our multivariate calibration software and implemented graphics capabilities in the software.

REFERENCES

1. G. L. McClure, ed., *Computerized Quantitative Infrared Analysis*, ASTM Special Technical Publication 934, Philadelphia (1987).
2. G. L. McClure, "Quantitative Analysis from the Infrared Spectrum," in: *Laboratory Methods in Vibrational Spectroscopy*, 3rd ed. (H. A. Willis, J. H. van der Maas, and R. G. J. Miller, eds.), pp. 145–201, Wiley, New York (1987).

3. D. M. Haaland, "Multivariate Calibration Methods Applied to Quantitative FT-IR Analyses," in: *Practical Fourier Transform Infrared Spectroscopy* (J. R. Ferraro and K. Krishnan, eds.), pp. 395–468, Academic Press, New York (1990).

4. D. M. Haaland, R. G. Easterling, and D. A. Vopicka, "Multivariate Least-Squares Methods Applied to the Quantitative Spectral Analysis of Multicomponent Samples," *Appl. Spectrosc.* **39**, 73–84 (1985).

5. H. J. Kisner, C. W. Brown, and G. J. Kavarnos, "Multiple Analytical Frequencies and Standards for the Least-Squares Spectrometric Analysis of Serum Lipids," *Anal. Chem.* **55**, 1703–1707 (1983).

6. G. L. McClure, P. B. Roush, J. F. Williams, and C. A. Lehmann, "Application of Computerized Quantitative Infrared Spectroscopy to the Determination of the Principal Lipids Found in Blood Serum," in: *Computerized Quantitative Infrared Analysis* (G. L. McClure, ed.), pp. 131–154, ASTM Special Publication 934, Philadelphia (1987).

7. C. K. Mann, J. R. Goleniewski, and C. A. Sismanidis, "Spectrophotometric Analysis by Cross-Correlation," *Appl. Spectrosc.* **36**, 223–228 (1982).

8. S. L. Monfre and S. D. Brown, "Estimation of Ester Hydrolysis Parameters by using FTIR Spectroscopy and the Extended Kalman Filter," *Anal. Chim. Acta* **200**, 397 (1988).

9. D. M. Haaland and E. V. Thomas, "Partial Least-Squares Methods for Spectral Analyses. 1. Relation to Other Quantitative Calibration Methods and the Extraction of Qualitative Information," *Anal. Chem.* **60**, 1193–1202 (1988).

10. D. M. Haaland and E. V. Thomas, "Partial Least-Squares Methods for Spectral Analyses. 2. Application to Simulated and Glass Spectral Data," *Anal. Chem.* **60**, 1202–1208 (1988).

11. D. M. Haaland, "Quantitative Infrared Analysis of Borophosphosilicate Films Using Multivariate Statistical Methods," *Anal. Chem.* **60**, 1208–1217 (1988).

12. M. P. Fuller, G. L. Ritter, and C. S. Draper, Partial Least-Squares Quantitative Analysis of Infrared Spectroscopic Data. Part I: Algorithm Implementation," *Appl. Spectrosc.* **42**, 217–227 (1988).

13. M. P. Fuller, G. L. Ritter, and C. S. Draper, "Partial Least-Squares Quantitative Analysis of Infrared Spectroscopic Data. Part II: Application to Detergent Analysis," *Appl. Spectrosc.* **42**, 228–236 (1988).

14. P. M. Fredericks, J. B. Lee, P. R. Osborn, and D. A. J. Swinkels, "Materials Characterization Using Factor Analysis of FT-IR Spectra. Part 1: Results," *Appl. Spectrosc.* **39**, 303–310 (1985).

15. P. M. Fredericks, J. B. Lee, P. R. Osborn, and D. A. J. Swinkels, "Materials Characterization Using Factor Analysis of FT-IR Spectra. Part 2: Mathematical and Statistical Considerations," *Appl. Spectrosc.* **39**, 311–316 (1985).

16. E. V. Thomas and D. M. Haaland, "Comparison of Multivariate Calibration Methods for Quantitative Spectral Analysis," *Anal. Chem.* **62**, 1091–1099 (1990).

17. D. M. Haaland, "Theoretical Comparison of Classical (K-Matrix) and Inverse (P-Matrix) Least-Squares Methods for Quantitative Infrared Spectroscopy," *SPIE* **553**, 241–242 (1985).

18. H. Martens and T. Naes, "Multivariate Calibration by Data Compression," in: *Near-infrared Technology in Agricultural and Food Industries* (P. C. Williams and K. Norris, eds.), pp. 57–87, Am. Assoc. Cereal Chem., St. Paul, Minnesota (1987).

19. G. E. P. Box, W. G. Hunter, and J. S. Hunter, *Statistics for Experimenters, An Introduction to Design, Data Analysis, and Model Building*, pp. 306–373, Wiley, New York (1978).

20. D. M. Haaland and R. G. Easterling, "Improved Sensitivity of Infrared Spectroscopy by the Application of Least Squares Methods," *Appl. Spectrosc.* **34**, 539–548 (1980).

21. M. Stone, "Cross-validatory Choice and Assessment of Statistical Predictions," *J. R. Stat. Soc., B* **36**, 111–133 (1974).

22. W. Lindberg, J.-A. Persson, and S. Wold, "Partial Least-Squares Method for Spectro-

fluorimetric Analysis of Mixtures of Humic Acid and Ligninsulfonate," *Anal. Chem.* **55**, 643–648 (1983).

23. S. Weisberg, *Applied Linear Regression*, pp. 106–125, Wiley, New York (1985).
24. D. M. Haaland and E. V. Thomas, "Comparison of Multivariate Quantitative Spectral Analysis Methods in the Presence of Nonlinearities," paper 1028, presented at the 1988 Pittsburgh Conference & Exposition on Analytical Chemistry and Applied Spectroscopy, February 22–26, 1988, New Orleans, Louisiana.
25. D. M. Haaland, "Partial Least-Squares Calibration Diagnostics Applied to the FT-IR Analysis of Borophosphosilicate Glass (BPSG) Thin Films," *7th International Conference on Fourier Transform Spectroscopy* (D. G. Cameron, ed.), *SPIE* **1145**, 425–426 (1989).

Hadamard Methods in Signal Recovery

2

Stephen A. Dyer, Robert M. Hammaker, and William G. Fateley

2.1. INTRODUCTION

Two well-established methods for performing optical spectrometry are dispersive spectrometry (DS) and Fourier transform spectrometry (FTS). A third alternative is Hadamard transform spectrometry (HTS), which is a marriage of dispersive and multiplexing techniques and possesses some of the desirable features of both DS and FTS.[1]

2.1.1. The Multiplex Advantage

Multiplexing is measuring the total energy (or, more accurately, the irradiance) of combinations of spectral resolution elements by having a multitude of frequencies (i.e., information from a multitude of spectral resolution elements) simultaneously impinging on a single detector. The purpose of multiplexing is to maximize the radiant flux incident on the single detector in order to increase the signal-to-noise ratio (SNR). Multiplexing will produce the desired increase in SNR only if the noise is independent of the strength of the incident radiation.[2]

Stephen A. Dyer • Department of Electrical and Computer Engineering, Kansas State University, Manhattan, Kansas 66506.   Robert M. Hammaker and William G. Fateley • Department of Chemistry, Kansas State University, Manhattan, Kansas 66506.

Computer-Enhanced Analytical Spectroscopy, Volume 3, edited by Peter C. Jurs. Plenum Press, New York, 1992.

Two established multiplexing techniques are: (1) interference techniques and Fourier transforms (normally performed with a Michelson interferometer), which yield Fourier transform (FT) spectrometers; and (2) encoding masks and Hadamard transforms, which yield Hadamard transform (HT) spectrometers.[3] If there is a dominant noise source, then its effect may be represented as $noise = (signal)^m$. For the desired case of $m = 0$ and for N spectral resolution elements, the SNR relative to that of DS is $\sqrt{N}$ for FTS and $\sqrt{N}/2$ for HTS for equal measurement time in all three instruments. This multiplex advantage for the favorable case of $m = 0$ arises because all N spectral resolution elements for FTS or $(N + 1)/2$ of the spectral resolution elements for HTS are incident on the single detector simultaneously.

An alternative to multiplexing is the use of a dispersive spectrometer with a multichannel detection system. The resulting multichannel advantage yields an improvement in SNR of $\sqrt{N}$, compared to a scanning dispersive spectrometer, for equal measurement time. This multichannel advantage arises because all N spectral resolution elements are incident on the N different detectors simultaneously.

The characteristics of the detection system determine the spectrometric method of choice if SNR is the primary concern. Three interesting possibilities for m in the relation $noise = (signal)^m$ are presented in Table 2.1. The multiplex advantage for $m = 0$ becomes multiplex neutrality for FTS and a multiplex disadvantage for HTS for $m = 1/2$ and a multiplex disadvantage for both FTS and HTS for $m = 1$, compared to a scanning dispersive spectrometer for equal measurement time. The multichannel advantage is present for all values of m at the cost of having N detectors rather than a single detector. Thus, when cost and availability are of no concern, a multichannel detection system can be an attractive alternative to any multiplexing technique. Our concern in this contribution is for situations in which multichannel detection is not practical and $m = 0$ in the relation $noise = (signal)^m$.

2.1.2. An Exemplary HTS System

The general features of an HT spectrometer are illustrated in Figs. 2.1 and 2.2. The Hadamard encoding mask is placed in a flat focal plane of a dispersive spectrometer. Figure 2.1a considers the case for $N = 7$ spectral resolution elements and shows the first of the 7 different encodements which must be measured to obtain the spectrum consisting of the 7 spectral resolution elements. In each encodement, $(N + 1)/2 = 4$ of the 7 spectral resolution elements pass through the mask to the detector and the other $(N - 1)/2 = 3$ spectral resolution elements are blocked by the mask. Here a movable mask consisting of $(2N - 1) = 13$ different elements is translated

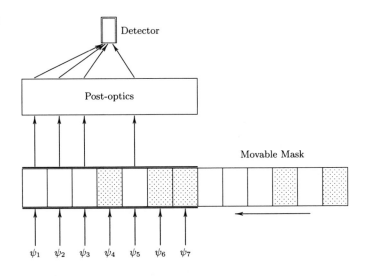

Intensity at Detector:

$$\eta_1 = (1)\psi_1 + (1)\psi_2 + (1)\psi_3 + (0)\psi_4 + (1)\psi_5 + (0)\psi_6 + (0)\psi_7 + e_1$$

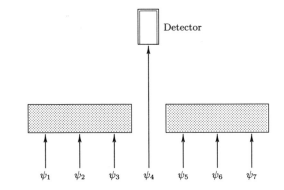

FIGURE 2.1. (a) The movable-mask arrangement in the focal plane for an HT spectrometer with $N = 7$ spectral resolution elements represented by seven arrows incident on the focal plane. The shaded areas of the movable mask are opaque to the radiation, whereas the white areas are transparent to the radiation. Here, four rays impinge on the detector. The intensity at the detector is composed of $\psi_1 + \psi_2 + \psi_3 + \psi_5$ plus the noise e_1. The movable mask is translated from right to left through seven different positions to provide the seven detector responses $\eta_1, \eta_2, ..., \eta_7$. (b) A single fixed exit slit of a dispersive spectrometer. Note that only one spectral element reaches the detector. The detector response is ψ_4 plus the noise.

TABLE 2.1. Effect of Detection System Characteristics
on the Choice of Spectrometric Method for the Best SNR

Detection system status	m	Best signal-to-noise ratio
Detector-noise limited	0	FTS or HTS
Photon-noise limited	$\frac{1}{2}$	?
Fluctuation-noise limited	1	DS

one element to the left each time to generate each of the 7 different combinations of 4 transparent and 3 opaque mask elements needed. The proper sequence of 13 transparent and opaque elements to use in constructing this movable Hadamard encoding mask is determined by Hadamard mathematics. Figure 2.1b represents a scanning dispersive spectrometer in which one spectral resolution element at a time is incident on the detector. Figure 2.2 shows the general relationship among the components of an HT spectrometer that contains a stationary Hadamard encoding mask rather than a movable one. As described later in Sections 2.1.3 and 2.2, our initial contribution to the field of HTS was the introduction of the stationary Hadamard encoding mask.[4,5] The specifics of our HT spectrometer are available in the literature.[5-7] The major optical problem is to recombine (i.e., dedisperse) the radiation that exits the mask into pseudowhite light of convenient image size to impinge on the single detector.

The problem of the most efficient way to carry out a multiplexing technique like HTS has been treated in general via weighting designs using matrix algebra.[8] The initial application was to balances, but optical instruments are completely analogous, as illustrated in Table 2.2. The case of interest is the optical instrument with a single detector and the corresponding S-matrix. (The S-matrix is a special case of the weighing matrix $\mathbf{W}$, described in more detail in Section 2.3.4.) For a movable mask such as the one illustrated in Fig. 2.1a, the S-matrix must be cyclic. Not all methods for generating S-matrices yield cyclic matrices, but three known

TABLE 2.2. Weighing Designs and Optical Multiplexing

Instruments and characteristics	Dual-detection	Single-detection
Balance	Double pan	Single pan
Optical instrument	Two detectors	Single detector
Weighing design matrix	Hadamard matrix	S-matrix
(w = integer)	$n \times n$	$N \times N$
	$n = 2^w$	$N = n - 1 = 2^w - 1$
Matrix elements	$(n/2)$ are $+1$	$(N+1)/2$ are $+1$
	$(n/2)$ are -1	$(N-1)/2$ are 0

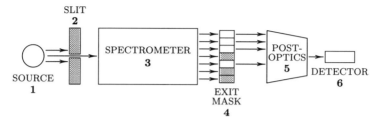

FIGURE 2.2. A dispersive spectrometer fitted with a stationary Hadamard encoding mask. The infrared source, 1, provides radiation to pass through the entrance slit, 2. The spectrometer, 3, disperses the radiation to a flat focal plane on the stationary exit mask, 4. The post-optics, 5, collect the radiation passing through the transparent elements and focus this radiation on the detector, 6.

constructions of cyclic S-matrices are available.[9] One of these is the quadratic residue construction, which for $N = 7$ gives the S-matrix

$$
\mathbf{S}_7 = \begin{bmatrix}
1 & 1 & 1 & 0 & 1 & 0 & 0 \\
1 & 1 & 0 & 1 & 0 & 0 & 1 \\
1 & 0 & 1 & 0 & 0 & 1 & 1 \\
0 & 1 & 0 & 0 & 1 & 1 & 1 \\
1 & 0 & 0 & 1 & 1 & 1 & 0 \\
0 & 0 & 1 & 1 & 1 & 0 & 1 \\
0 & 1 & 1 & 1 & 0 & 1 & 0
\end{bmatrix}
$$

for Fig. 2.1a. Note that only the first row of a cyclic S-matrix is needed in order to construct all remaining rows. Each subsequent row is obtained from the previous row by a left-shift registration (in which the element in the far left column moves to the far right column of the next row and each of the other $N-1$ elements moves one column to the left).

TABLE 2.3. Section of a Simulated
Near-Infrared Spectrum

i	ψ_i (arbitrary units)	ν_i (cm^{-1})
1	9	10,100
2	6	10,200
3	3	10,300
4	4	10,400
5	1	10,500
6	6	10,600
7	10	10,700

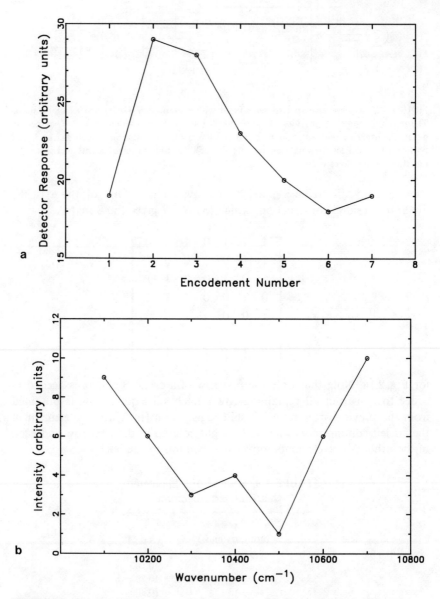

FIGURE 2.3. The encodegram (a) and the spectrum (b) for a simulated NIR spectrum consisting of $N = 7$ spectral resolution elements.

To illustrate the operation of the HT spectrometer presented schematically in Fig. 2.1.a or Fig. 2.2 we consider the section of a simulated near-infrared (NIR) spectrum having seven energies ψ_i, $i = 1, ..., 7$, at seven frequencies v_i, $i = 1, ..., 7$, listed in Table 2.3. The seven rows of S_7 are the seven different encodements needed to obtain the spectrum, and the seven detector responses η_i, $i = 1, ..., 7$, are as follows:

$$\eta_1 = (1)\psi_1 + (1)\psi_2 + (1)\psi_3 + (0)\psi_4 + (1)\psi_5 + (0)\psi_6 + (0)\psi_7 = 19$$

$$\eta_2 = (1)\psi_1 + (1)\psi_2 + (0)\psi_3 + (1)\psi_4 + (0)\psi_5 + (0)\psi_6 + (1)\psi_7 = 29$$

$$\eta_3 = (1)\psi_1 + (0)\psi_2 + (1)\psi_3 + (0)\psi_4 + (0)\psi_5 + (1)\psi_6 + (1)\psi_7 = 28$$

$$\eta_4 = (0)\psi_1 + (1)\psi_2 + (0)\psi_3 + (0)\psi_4 + (1)\psi_5 + (1)\psi_6 + (1)\psi_7 = 23$$

$$\eta_5 = (1)\psi_1 + (0)\psi_2 + (0)\psi_3 + (1)\psi_4 + (1)\psi_5 + (1)\psi_6 + (0)\psi_7 = 20$$

$$\eta_6 = (0)\psi_1 + (0)\psi_2 + (1)\psi_3 + (1)\psi_4 + (1)\psi_5 + (0)\psi_6 + (1)\psi_7 = 18$$

$$\eta_7 = (0)\psi_1 + (1)\psi_2 + (1)\psi_3 + (1)\psi_4 + (0)\psi_5 + (1)\psi_6 + (0)\psi_7 = 19$$

Each detector response η_i also contains an error term e_i due to noise. This error term has been omitted from the equations for η_i for simplicity.

The primary data from an HT spectrometer is a plot of these detector responses versus encodement number as shown in Fig. 2.3a (where the points for the seven different detector responses are connected by straight lines). This plot is called an encodegram in analogy to the interferogram as a plot of detector response versus optical retardation or time for the Michelson interferometer in an FT spectrometer. The vector-matrix equation describing the generation of the encodegram in Fig. 2.3a is

$$\boldsymbol{\eta} = S_7 \boldsymbol{\psi}$$

where $\boldsymbol{\eta} = [\eta_i]$ and $\boldsymbol{\psi} = [\psi_i]$ are column vectors of seven rows. The spectrum $\boldsymbol{\psi}$ of Fig. 2.3b (where the points for the seven different (v_i, ψ_i)-pairs are connected by straight lines) can be recovered from the encodegram of Fig. 2.3a through the vector-matrix equation

$$\boldsymbol{\psi} = S_7^{-1} \boldsymbol{\eta} \qquad (2.1)$$

An efficient indirect means of solving Eq. (2.1) is through the use of a fast Hadamard transform (FHT).[10,11] Figure 2.4 shows an encodegram/spectrum pair obtained from the Raman scattering of cyclohexane excited by the 514.5 nm line of an Ar^+ laser and measured using our visible HT-Raman spectrometer.[6,7,12]

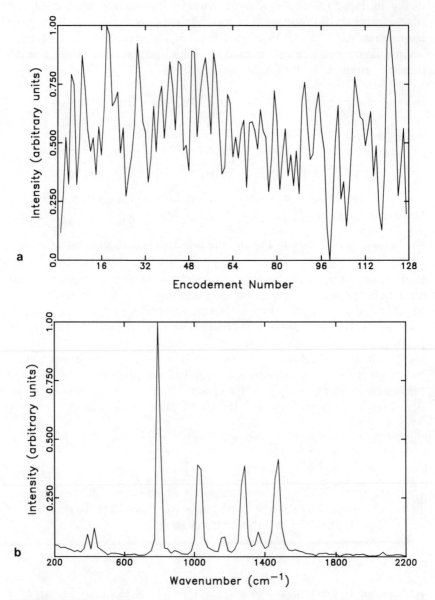

FIGURE 2.4. The encodegram (a) and the spectrum (b) for the Raman scattering of cyclohexane excited by the 514.5 nm line of an Ar⁺ laser.

The HT spectrometer developed by Decker in the early 1970s employed a movable mask.[13] Significant problems such as misalignment and jamming associated with a movable mask contributed to a dormant period in the development of HTS. Our interest in HTS began in the late 1970s and our original contribution was the introduction of the stationary Hadamard encoding mask in the 1980s.[4,5] The use of electrooptical materials to fabricate the mask elements provides a stationary Hadamard encoding mask and eliminates the problems associated with continuously moving parts. In addition, for N spectral resolution elements, the stationary mask requires N elements, whereas the movable mask requires $2N - 1$ elements. The advantages associated with the electrooptical masks do not come without a price.[14] These masks have nonideal transmittance characteristics: they are not totally transmissive in the "on" state, they are not completely opaque in the "off" state, and their transmittances are wavelength-dependent. Thus, we trade the mechanical problems of the movable mask, whose elements are totally transmissive or completely opaque, for the optical problems of the stationary mask. However, if the transmittances of a mask are known, procedures can be used to compensate for the non-idealities. The major concern of this contribution is to find the optimal compensation scheme, and achieving that goal is the subject of Sections 2.3 and 2.4.

2.1.3. Comparison with FTS

The advantages of FTS compared to DS are: (1) the multiplex, or Fellgett, advantage[15–20]; (2) the frequency-precision, or Connes, advantage; and (3) the throughput, or Jacquinot, advantage.[18,20–23] As already noted in Section 2.1.1, the Fellgett advantage arises from the detection of a multitude of frequencies (i.e., information from a multitude of spectral resolution elements) simultaneously impinging on a single detector. The Connes advantage arises from the laser measurement of the moving-mirror position, or optical retardation, resulting in good co-addition of interferograms. Precision in measuring optical retardation transforms into precision in frequency in spectra, resulting in good spectral subtraction. The Jacquinot advantage arises from the fact that an interferometer uses apertures, while a dispersive spectrometer uses slits. The solid angle of the collimated beam passing through the interferometer is greater than that for a scanning dispersive spectrometer operating at the same resolution.

The HT spectrometer possesses a multiplex advantage as noted in Section 2.1.1. When a stationary Hadamard encoding mask is used in an HT spectrometer, the entire instrument has no continuously moving parts. Consequently, the spectral character of a spectral resolution element passing through a given mask element remains unchanged during the experiment. This feature provides an advantage analogous to the Connes advantage in FTS. Our HT spectrometer exhibits good co-addition and spectral subtraction in practice.[5,6,7,12,24] Since dispersive spectrometers generally have f-numbers considerably larger (>4) than those of interferometers (1–4), no Jacquinot advantage is expected for HT spectrometers.

Many samples, including polymers and biological compounds, cannot be studied by conventional visible Raman spectrometry due to laser-induced fluorescence, which generally overwhelms the weak Raman scattering and yields only a fluorescence signal. The use of lasers that operate at longer wavelength (e.g., HeNe at 632.8 nm, Kr^+ at 647.1 nm, GaAlAs diode at 783 nm and Nd:YAG at 1064 nm), for which the laser photons lack the energy necessary to excite fluorescence, can overcome this limitation. Since the intensity of Raman scattering is inversely proportional to the fourth power of the excitation wavelength, going from the midvisible (488.0 nm, 514.5 nm) spectral region to the NIR (1064 nm) spectral region reduces the already-small intensity of Raman scattering by up to a factor of 23. Consequently, multiplex spectrometry is an attractive method for attempting to counter the loss of signal resulting from the move from the midvisible spectral region to the NIR spectral region. The development of NIR FT-Raman spectrometry[25–28] began in the mid-1980s and prompted us to begin the development of NIR HT-Raman spectrometry.

Our initial feasibility studies were in visible HT-Raman spectrometry with the 514.5 nm line of the Ar^+ laser.[6,7,12,29] We then advanced to the more demanding NIR spectral region of interest, using the 1064 nm line of the Nd:YAG laser.[24] A major concern in FT-Raman spectrometry is the removal of all Rayleigh scattering by optical filtering before the Raman-scattered radiation is allowed to enter the interferometer. Any noise associated with the Rayleigh line will be distributed throughout the transformed spectrum and may drastically degrade the SNR.[30] Since all frequencies travel together through the interferometer, a choice between optical filtering and spatial filtering does not exist. However, since an HT spectrometer is a dispersive spectrometer, spatial filtering occurs naturally by positioning the grating so that the Rayleigh line is never incident on the mask. The effect of failure to address this concern can be demonstrated by intentional reduction in optical filtering in FT-Raman spectrometry[30] and intentional incorrect positioning of the grating in HT-Raman spectrometry.[31] A general comparison of HT and FT instruments is given in Table 2.4.

Inspection of Table 2.4 suggests that for simplicity and cost the clear choice is an HT-Raman spectrometer, but for overall capability the clear choice is an FT-Raman spectrometer. A significant advantage of the HT-Raman spectrometer is the use of spatial filtering for Rayleigh-line rejection.

TABLE 2.4. Instrumentation: A Comparison

Characteristic	HT-Raman	FT-Raman
Spectrometer	Dispersive	Interferometric
Spectral collection	Optical encodement via grating and mask	Moving mirror (after filtering)
Rayleigh-line rejection	Field-stop mask (spatial filtering)	Optical filtering
Transformation	Fast HT	Fast FT
Multiplexing (Fellgett advantage)	Yes	Yes
Jacquinot advantage	No	Yes
Co-addition/subtraction (Connes advantage)	Yes	Yes
Resolution	Low	High
Spectral range	Narrow	Wide
Dynamic range	Small	Large
Simple electronics and computing	Yes	No
Continuously moving parts	No	Yes

A great advantage of the FT-Raman spectrometer is the possibility of improving resolution without sacrificing spectral range by increasing the extent of mirror travel in the moving-mirror arm of the Michelson interferometer.

For an HT spectrometer with a mask having a given number of elements of a given width, the dispersion of the grating determines the spectral range incident on the mask and the size of the spectral resolution elements. Increasing the dispersion of the grating will improve the resolution by making the size of the mask element and corresponding spectral resolution element correspond to a smaller range of frequencies. The resulting cost is a decrease in the spectral range incident on the mask.[32] Thus, the major potential limitations of HT spectrometers are low resolution and limited spectral range for a given grating position. In many cases these limitations are counterbalanced by the simplicity of the instrumentation.

The trade of mechanical problems of the movable mask for the optical problems of the stationary mask contributes to the simplicity of the HT spectrometer. In addition to the Fellgett advantage and the analogy to the Connes advantage, HT spectrometers with stationary masks may possess the following features not necessarily enjoyed by other techniques:

1. Simplicity of no continuously moving parts.
2. A less complex amplifier and analog-to-digital converter associated with the detector, due to a smaller dynamic range of signal.

3.   Simpler computational needs.
4.   Ruggedness for use in hostile environments.
5.   Simpler maintenance and training requirements.
6.   Use of spatial filtering rather than optical filtering.
7.   Lower cost.

In our view, the future of HT spectrometers lies in applications in which many repetitive analyses in well-defined situations must be done in a cost-effective manner. Areas such as environmental monitoring and production-process stream monitoring and control immediately come to mind.

2.2. ELECTROOPTIC SHUTTERS

Three possible types of materials for use as mask elements operating as optical shutters are: liquid crystals,[5] thermodiachromatic crystals (changing transmission properties via semiconductor/metal transitions),[4] and electrochromic materials such as WO_3. Our work has been restricted to liquid-crystal materials.

Initially, we modified a commercially available liquid-crystal optical shutter array (LC-OSA) to meet our needs.[5] The resulting stationary Hadamard encoding mask, possessing 127 elements, uses a polarizer-analyzer pair to generate the opaque and transmissive states. The total transmittance of any element in the LC-OSA can never be greater than 0.5 for randomly polarized radiation, due to the limitation of linear polarizers.[6, 33] In addition, the "off" state of the LC-OSA is not completely opaque. Nevertheless, the LC-OSA performed well in a visible HT-Raman spectrometer.[6, 7, 12, 29] Conversion to an NIR HT-Raman spectrometer was accomplished by replacing the visible polarizers with NIR glass polarizers. These NIR glass polarizers are not as satisfactory as the visible polarizers. However, even with an LC-OSA having poor transmission properties it is possible to obtain satisfactory NIR HT-Raman spectra.[24]

Recently, we found liquid-crystal materials that do not require a polarizer-analyzer pair.[34] Instead of relying on polarization phenomena, as our initial LC-OSA does, these materials act as Mie scatters when in the "off" state. There are four liquid-crystal materials available to us that work on the principle of scattering and transmission in their "off" and "on" states, respectively. One is called a colloidal-dispersed liquid crystal (CDLC) and three are called polymer-dispersed liquid crystals (PDLCs). The four materials vary in optical properties, switching times, and ease of fabrication. Based on preliminary work with masks fabricated from the CDLC material and two of the three PDLC formulations, we have chosen one of the PDLC formulations for development as the second generation of stationary Hadamard encoding masks.

Figure 2.5 presents the NIR spectra (10,000–5500 cm^{-1}) of our original LC-OSA (dashed lines) and a bulk sample of the PDLC material used in the three different formulations of PDLC mask materials (solid lines). Our second-generation stationary Hadamard encoding mask possesses 255 elements, each of which has transmission properties similar to those of the bulk PDLC material in Fig. 2.5. The overall transmission of the PDLC mask is, of course, reduced by the absorption of the spacers separating the mask elements. An analysis of a uniformly imperfect stationary Hadamard encoding mask shows that the SNR is approximately proportional to $\Delta\tau$, the difference between the transmittance τ_t in the transmissive state and the transmittance τ_o in the opaque state, where $\Delta\tau = 1 - 0 = 1$ for a perfect mask.[14] Figure 2.5 shows that, for the LC-OSA, $\Delta\tau$ varies from 0.12 to 0.10 for 9400–8300 cm^{-1}, corresponding to a Stokes Raman shift of up to 1100 cm^{-1} from the 1064 nm line of the Nd:YAG laser. For the mask fabricated using the PDLC material shown in Fig. 2.5, $\Delta\tau$ varies from 0.55 to 0.45 for 10,000–5500 cm^{-1}, corresponding to Stokes Raman shifts of up to 3900 cm^{-1} from the 1064 nm line of the Nd:YAG laser. The advance needed for NIR HT-Raman spectrometry in particular and HTS in general is the development of signal recovery methods to compensate for non-idealities of the mask.

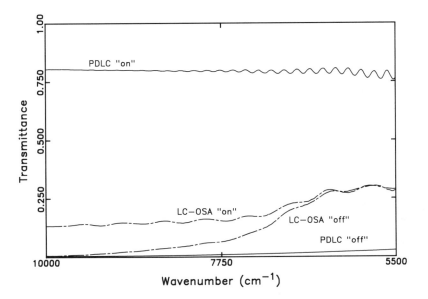

FIGURE 2.5. Transmission spectra of the liquid-crystal optical shutter array and a bulk sample of the polymer-dispersed liquid-crystal material.

2.3. RECOVERY OF SPECTRA

Figure 2.6 provides a basic block diagram of the operation of a multiplexing spectrometer.

The actual spectrum to be measured is represented by

$$\boldsymbol{\psi} = [\psi_1 \psi_2 \cdots \psi_N]^{\mathrm{T}}$$

where ψ_i is the energy (or, more accurately, the spectral irradiance) incident on the ith element of an N-element encoding mask (i.e., the ith spectral resolution element), and T denotes "transpose."

Quantity $\tilde{\mathbf{W}}$ is an N-square matrix, described in more detail later, which represents the encoding mask and the weighing design used.

The detector-error vector

$$\mathbf{e} = [e_1 e_2 \cdots e_N]^{\mathrm{T}}$$

contains the errors, due to detector noise, associated with the N measurements taken.

The N measurements made are represented by the vector

$$\boldsymbol{\eta} = [\eta_1 \eta_2 \cdots \eta_N]^{\mathrm{T}}$$

where

$$\boldsymbol{\eta} = \tilde{\mathbf{W}}\boldsymbol{\psi} + \mathbf{e} \qquad (2.2)$$

Once $\boldsymbol{\eta}$ is available, some operation must be performed on it to obtain an estimate $\hat{\boldsymbol{\psi}} = [\hat{\psi}_1 \hat{\psi}_2 \cdots \hat{\psi}_N]^{\mathrm{T}}$ of the true spectrum $\boldsymbol{\psi}$. It is generally desirable to have $\hat{\boldsymbol{\psi}}$ be "good" in some sense.

One condition typically placed on the estimate is that it be unbiased; that is,

$$\mathrm{E}[\hat{\boldsymbol{\psi}}] = \boldsymbol{\psi} \qquad (2.3)$$

where $\mathrm{E}[\cdot]$ denotes the operation of expectation. This property says that the estimate gives, on the average, the true spectrum.

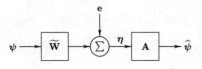

FIGURE 2.6. Basic block diagram of a multiplexing spectrometer.

2.3.1. Assumptions on the Error

The following assumptions are made as we proceed through the development of the spectrum-recovery methods:

1. The encoding mask has elements of width sufficient enough to allow one to neglect the effects of diffraction.
2. The error e_j associated with a given measurement η_j is a random variable and is independent of the amount of radiation impinging on the detector.
3. The error is zero-mean; that is,

$$E[e_j] = 0$$

4. The errors in distinct measurements are uncorrelated; that is,

$$E[e_j e_k] = \begin{cases} \sigma^2 & j=k \\ 0, & j \neq k \end{cases}$$

where σ^2 denotes the variance of e_j.

2.3.2. Coaddition

Coaddition is a technique which consists of averaging several random variables to form a new random variable having reduced variance. Consider L independent, identically distributed random variables x_i, $i = 1, ..., L$, each with mean μ_x and variance σ_x^2. A new random variable y can be defined as

$$y = \frac{1}{L} \sum_{i=1}^{L} x_i$$

The mean μ_y and the variance σ_y^2 of y are

$$\mu_y = \mu_x$$

and

$$\sigma_y^2 = \frac{\sigma_x^2}{L} \tag{2.4}$$

respectively.

2.3.3. The Optimum Unbiased Linear Spectrum-Estimate

Another restriction generally placed on $\hat{\psi}$ is that it be a linear function of the measurements; that is,

$$\hat{\psi} = A\eta \tag{2.5}$$

where A is an N-square matrix.

We can rewrite Eq. (2.5), using Eq. (2.2), to get

$$\hat{\psi} = A\tilde{W}\psi + Ae \tag{2.6}$$

For Eq. (2.3) to hold, we must have

$$A = \tilde{W}^{-1} \tag{2.7}$$

It turns out[35] that, if the detector noise is Gaussian, the estimate

$$\hat{\psi} = \tilde{W}^{-1}\eta \tag{2.8}$$

is the maximum-likelihood estimate, which in turn implies that $\hat{\psi}$ is the optimal unbiased linear estimate.

The vector of errors in the estimates is

$$\varepsilon = \hat{\psi} - \psi$$
$$= \tilde{W}^{-1}e$$

That is, the error in the jth estimate is

$$\varepsilon_j = \hat{\psi}_j - \psi_j$$
$$= \sum_{i=1}^{N} \xi_{ji} e_i$$

where ξ_{ji} denotes the jith element of $\tilde{W}^{-1}$. The mean-square error (MSE) in the jth estimate is

$$\epsilon_j = E[\varepsilon_j^2]$$

The average MSE (AMSE) associated with the estimate of Eq. (2.8) is

$$\epsilon = \frac{1}{N} \sum_{j=1}^{N} \epsilon_j$$
$$= \frac{\sigma^2}{N} \operatorname{tr}\{(\tilde{\mathbf{W}}^T \tilde{\mathbf{W}})^{-1}\} \tag{2.9}$$

where $\operatorname{tr}\{\cdot\}$ denotes the trace.

As a simple example, we take a conventional monochromator. If the monochromator is ideal, then the weighing matrix associated with it is

$$\tilde{\mathbf{W}} = \mathbf{I}$$

where $\mathbf{I}$ is the N-square identity matrix. The AMSE is, from Eq. (2.9),

$$\epsilon^{(I)} = \frac{\sigma^2}{N} \operatorname{tr}\{(\mathbf{I}^T \mathbf{I})^{-1}\}$$
$$= \sigma^2$$

as expected.

2.3.4. Representation of Nonideal Masks[36,37]

We represent a nonideal mask, which is used to obtain N measurements and which has a total of N elements, by the N-square matrix

$$\tilde{\mathbf{W}} = [\tilde{w}_{ij}]$$

where i indexes the measurement, and j indexes the spectral resolution element.

The elements $\tilde{w}_{ij}$ take on values as follows:

$$\tilde{w}_{ij} = \begin{cases} \tau_{oj}, & \text{if the } j\text{th mask element is ``off''} \\ & \text{(opaque) during the } i\text{th measurement} \\ \tau_{tj}, & \text{if the } j\text{th mask element is ``on''} \\ & \text{(transmissive) during the } i\text{th measurement} \end{cases}$$

The transmittances τ_{oj} and $\tau_{tj}, j = 1, ..., N$, are assumed to be known, having been determined by previous characterization of the mask.

Our desire is to rewrite $\tilde{\mathbf{W}}$ as

$$\tilde{\mathbf{W}} = \mathbf{W}\mathbf{T} \tag{2.10}$$

where $\mathbf{W}$, called the weighing matrix and consisting only of ones and zeros, represents the weighing design (i.e., the measurement scheme) employed, and $\mathbf{T}$ is a transfer matrix which incorporates the description of the mask. If the mask were ideal, then $\mathbf{T}$ would equal the identity matrix $\mathbf{I}$.

Alternatively, we can write $\tilde{\mathbf{W}}$ as

$$\tilde{\mathbf{W}} = \tilde{\mathbf{D}} + \tilde{\mathbf{T}}_o \tag{2.11}$$

where

$$\tilde{\mathbf{D}} = [\tilde{d}_{ij}]$$

with

$$\tilde{d}_{ij} = \tilde{w}_{ij} - \tau_{oj}$$

$$= \begin{cases} 0, & \text{if the } j\text{th mask element is "off"} \\ & \text{during the } i\text{th measurement} \\ \tau_{tj} - \tau_{oj}, & \text{if the } j\text{th mask element is "on"} \\ & \text{during the } i\text{th measurement} \end{cases}$$

and

$$\tilde{\mathbf{T}}_o = \begin{bmatrix} \tau_{o1} & \tau_{o2} & \cdots & \tau_{oN} \\ \tau_{o1} & \tau_{o2} & \cdots & \tau_{oN} \\ \vdots & \vdots & \ddots & \vdots \\ \tau_{o1} & \tau_{o2} & \cdots & \tau_{oN} \end{bmatrix}$$

We now define the diagonal matrices $\mathbf{T}_o$, $\mathbf{T}_t$, and $\mathbf{D}$ as follows:

$$\mathbf{T}_o = \text{diag}(\tau_{o1}, \tau_{o2}, ..., \tau_{oN})$$

$$\mathbf{T}_t = \text{diag}(\tau_{t1}, \tau_{t2}, ..., \tau_{tN})$$

and

$$\mathbf{D} = \text{diag}(d_{11}, d_{22}, ..., d_{NN})$$

$$= \mathbf{T}_t - \mathbf{T}_o$$

Then $\tilde{\mathbf{D}}$ can be written as

$$\tilde{\mathbf{D}} = \mathbf{W}\mathbf{D} \tag{2.12}$$

and $\tilde{\mathbf{T}}_o$ can be written as

$$\tilde{\mathbf{T}}_o = \mathbf{J}\mathbf{T}_o \tag{2.13}$$

where $\mathbf{J}$ is an N-square matrix of ones. Alternatively, we can write Eq. (2.13) as

$$\tilde{\mathbf{T}}_o = \mathbf{1}\mathbf{\tau}_o^T \tag{2.14}$$

where

$$\mathbf{1} = [\, 11 \cdots 1\,]^T$$

and

$$\mathbf{\tau}_o = [\, \tau_{o1}\, \tau_{o2} \cdots \tau_{oN}\,]^T$$

Using Eqs. (2.12) and (2.14), we can rewrite Eq. (2.11) as

$$\tilde{\mathbf{W}} = \mathbf{W}\mathbf{D} + \mathbf{1}\mathbf{\tau}_o^T \tag{2.15}$$

In order for $\mathbf{T}$ to be found, Eq. (2.15) can be rewritten as

$$\tilde{\mathbf{W}} = \mathbf{W}(\mathbf{D} + \mathbf{\omega}\mathbf{\tau}_o^T) \tag{2.16}$$

where

$$\mathbf{\omega} = \mathbf{W}^{-1}\mathbf{1} \tag{2.17}$$

Comparing Eqs. (2.10) and (2.16), we see that the transfer matrix is

$$\mathbf{T} = \mathbf{D} + \mathbf{\omega}\mathbf{\tau}_o^T \tag{2.18}$$

Example. To illustrate, we consider a singly encoded HT spectrometer that uses the weighing design given by

$$\mathbf{W} = \begin{bmatrix} 1 & 0 & 1 \\ 0 & 1 & 1 \\ 1 & 1 & 0 \end{bmatrix} \tag{2.19}$$

A nonideal mask used in conjunction with the weighing design of Eq. (2.19) has the general representation given by

$$\tilde{\mathbf{W}} = \begin{bmatrix} \tau_{t1} & \tau_{o2} & \tau_{t3} \\ \tau_{o1} & \tau_{t2} & \tau_{t3} \\ \tau_{t1} & \tau_{t2} & \tau_{o3} \end{bmatrix}$$

According to Eq. (2.15),

$$\tilde{\mathbf{W}} = \mathbf{W}\mathbf{D} + \mathbf{1}\tau_o^T$$

$$= \begin{bmatrix} 1 & 0 & 1 \\ 0 & 1 & 1 \\ 1 & 1 & 0 \end{bmatrix} \begin{bmatrix} \tau_{t1} - \tau_{o1} & 0 & 0 \\ 0 & \tau_{t2} - \tau_{o2} & 0 \\ 0 & 0 & \tau_{t3} - \tau_{o3} \end{bmatrix} + \begin{bmatrix} 1 \\ 1 \\ 1 \end{bmatrix} \begin{bmatrix} \tau_{o1} \tau_{o2} \tau_{o3} \end{bmatrix}$$

$$= \begin{bmatrix} \tau_{t1} - \tau_{o1} & 0 & \tau_{t3} - \tau_{o3} \\ 0 & \tau_{t2} - \tau_{o2} & \tau_{t3} - \tau_{o3} \\ \tau_{t1} - \tau_{o1} & \tau_{t2} - \tau_{o2} & 0 \end{bmatrix} + \begin{bmatrix} \tau_{o1} & \tau_{o2} & \tau_{o3} \\ \tau_{o1} & \tau_{o2} & \tau_{o3} \\ \tau_{o1} & \tau_{o2} & \tau_{o3} \end{bmatrix}$$

$$= \begin{bmatrix} \tau_{t1} & \tau_{o2} & \tau_{t3} \\ \tau_{o1} & \tau_{t2} & \tau_{t3} \\ \tau_{t1} & \tau_{t2} & \tau_{o3} \end{bmatrix}$$

2.4. EFFICIENT METHODS FOR SIGNAL RECOVERY

2.4.1. The S-Method[38,39]

The best possible weighing design for a singly encoded HT spectrometer having a perfect mask is given by[40]

$$\mathbf{W} = \mathbf{S}$$

where $\mathbf{S}$ is an N-square simplex matrix.

For an ideal mask, the spectrum-recovery method to be used sets

$$\tilde{\mathbf{W}}^{-1} = \mathbf{S}^{-1} \tag{2.20}$$

giving

$$\hat{\psi} = \mathbf{S}^{-1}\eta \tag{2.21}$$

As mentioned in Section 2.1.2, the estimate $\hat{\psi}$ can be computed via an FHT.

The principal disadvantage in using the S-method is that any non-idealities present in the mask are ignored. If $\tilde{W} = ST$ and Eq. (2.21) is used to recover the spectrum, we get as the estimate

$$\hat{\psi} = S^{-1}[ST\psi + e]$$
$$= T\psi + S^{-1}e \qquad (2.22)$$

The estimate given by Eq. (2.22) is biased unless $T = I$, since $E[\hat{\psi}] \neq \psi$. Using Eq. (2.9), we can write the AMSE as

$$\epsilon = \frac{4N\sigma^2}{(N+1)^2} + \frac{1}{N}\operatorname{tr}\{(T-I)\,\psi\psi^T(T-I)^T\} \qquad (2.23)$$

Note that the second term in Eq. (2.23), the systematic error, is dependent on both the spectrum and the mask. This systematic error may contribute negligibly to the global error but significantly to the local error.

Equation (2.23) can, after some manipulation, be rewritten as

$$\epsilon = \frac{4N\sigma^2}{(N+1)^2} + \frac{1}{N}\sum_{i=1}^{N}\left[(d_{ii}-1)\,\psi_i + \frac{2}{N+1}\sum_{j=1}^{N}\tau_{oj}\psi_j\right]^2 \qquad (2.24)$$

The second term within the square brackets in Eq. (2.24) is constant for all i, and it is approximately twice as large as the average value of $\tau_{oj}\psi_j$.

In the ideal case, the AMSE associated with the (optimal unbiased) linear estimate $\hat{\psi}$ is, from either Eqs. (2.9) and (2.20) or Eq. (2.24),

$$\epsilon = \frac{4N\sigma^2}{(N+1)^2} \qquad (2.25)$$

In general, when L coadditions are used, the AMSE is

$$\epsilon(L) = \frac{4N\sigma^2}{L(N+1)^2} + \frac{1}{N}\sum_{i=1}^{N}\left[(d_{ii}-1)\,\psi_i + \frac{2}{N+1}\sum_{j=1}^{N}\tau_{oj}\psi_j\right]^2 \qquad (2.26)$$

The AMSE decreases as the number of coadditions is increased, although the systematic error is not affected. For large L, the AMSE is virtually equal to the systematic error, which reveals the bias of the S-method.

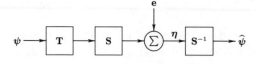

FIGURE 2.7. Block diagram of a multiplexing spectrometer that employs the S-method.

Procedure.   The S-method for a singly encoded HT spectrometer is carried out as follows. (Refer to Fig. 2.7.)

1. Perform the intended N spectrometric measurements, obtaining the measurement vector η.
2. Obtain the spectrum-estimate via the FHT as

$$\hat{\psi} = S^{-1}\eta$$

2.4.2. The D-Method

One computationally inexpensive method for obtaining a nearly optimal estimate $\hat{\psi}$ is detailed by Dyer et al.[36] We refer to this method as the D-method. Figure 2.8 makes use of Fig. 2.6 and Eq. (2.10) to provide a revised block diagram of the operation of a multiplexing spectrometer in which the recovery of the spectrum is optimal.

We assume that fast algorithms exist for obtaining

$$\hat{\psi}' = W^{-1}\eta \qquad (2.27)$$

for the weighing designs of interest. However, obtaining

$$\hat{\psi} = T^{-1}\hat{\psi}' \qquad (2.28)$$

by conventional means requires N^2 multiplications and $N(N-1)$ additions. In addition, T^{-1} must be found—although only once for a given spectrometer. Use of a direct method to find T^{-1} requires $O(N^3)$ operations, where $O(\cdot)$ denotes "on the order of."

An alternative approach can be developed by using Eq. (2.15) to rewrite the vector $\tilde{W}\psi$ of signals applied to the detector. That is,

$$\tilde{W}\psi = WD\psi + b \qquad (2.29)$$

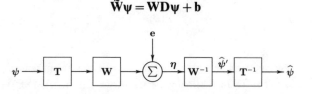

FIGURE 2.8. Revised block diagram of a multiplexing spectrometer.

where

$$\mathbf{b} = \mathbf{JT}_o\mathbf{\psi} = \mathbf{1}\mathbf{\tau}_o^T\mathbf{\psi}$$

Note that $\mathbf{b}$ is a constant vector, the value of each component being

$$b = \sum_{j=1}^{N} \tau_{oj}\psi_j. \tag{2.30}$$

The measurement vector is then

$$\mathbf{\eta} = \tilde{\mathbf{W}}\mathbf{\psi} + \mathbf{e}$$
$$= \mathbf{WD}\mathbf{\psi} + \mathbf{b} + \mathbf{e} \tag{2.31}$$

An estimate $\hat{\mathbf{\psi}}$ could be found as

$$\hat{\mathbf{\psi}} = \mathbf{D}^{-1}\mathbf{W}^{-1}\mathbf{\eta}' \tag{2.32}$$

where

$$\mathbf{\eta}' = \mathbf{\eta} - \mathbf{b}$$

Since $\mathbf{D}$ is diagonal, $\mathbf{D}^{-1}$ is also diagonal; namely,

$$\mathbf{D}^{-1} = \text{diag}(1/d_{11}, 1/d_{22}, ..., 1/d_{NN})$$

Once $\mathbf{W}^{-1}\mathbf{\eta}'$ is determined, only N additional multiplications are required in order to find $\hat{\mathbf{\psi}}$ as given in Eq. (2.32). Figure 2.9 shows a block diagram of the model that relies on Eqs. (2.29), (2.31), and (2.32).

To apply the D-method we need to obtain a good estimate $\hat{\mathbf{b}}$ of the offset vector $\mathbf{b}$. Since $\mathbf{b}$ is a constant vector, we need to find only one value, given by Eq. (2.30). An estimate $\hat{b}$ of the value b can be obtained by

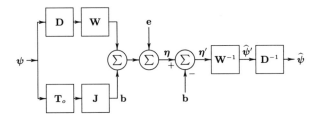

FIGURE 2.9. Model of a multiplexing spectrometer, which shows the basis of the D-method.

inserting the sample whose spectrum is to be determined, turning all mask elements "off," and then taking a measurement. That estimate is

$$\hat{b} = b + e_0 \tag{2.33}$$

where the error e_0 has the same statistics as those of the e_j, $j = 1, ..., N$.

For a single measurement, the MSE in the estimate $\hat{b}$ is

$$\epsilon_b(1) = \sigma^2$$

The MSE can be reduced to any desired level simply by averaging N_0 measurements of b. When N_0 measurements are averaged, the MSE in $\hat{b}$ is

$$\epsilon_b(N_0) = \frac{\sigma^2}{N_0}$$

The AMSE for the D-method is

$$\epsilon = \frac{\sigma^2}{N} \left(1 + \frac{1}{N_0}\right) \text{tr}\{\mathbf{D}^{-1}\mathbf{W}^{-1}(\mathbf{W}^{-1})^{\text{T}}\mathbf{D}^{-1}\} \tag{2.34}$$

It should be noted from Eq. (2.34) that relatively few measurements of b are required in practice. For example, taking only $N_0 = 10$ measurements results in the AMSE being just 10% above the minimum obtainable for a given mask.

For the case in which $\mathbf{W} = \mathbf{S}$, Eq. (2.34) becomes, after some manipulation,

$$\epsilon = \frac{4\sigma^2}{N(N+1)^2} \left(1 + \frac{1}{N_0}\right) \text{tr}\{\mathbf{D}^{-1}[(N+1)\mathbf{I} - \mathbf{J}]\mathbf{D}^{-1}\}$$

which in turn yields

$$\epsilon = \frac{4\sigma^2}{(N+1)^2} \left(1 + \frac{1}{N_0}\right) \sum_{i=1}^{N} \frac{1}{d_{ii}^2} \tag{2.35}$$

With L coadditions the AMSE is[38]

$$\epsilon(L) = \frac{4\sigma^2}{(N+1)^2} \left(\frac{1}{L} + \frac{1}{N_0}\right) \sum_{i=1}^{N} \frac{1}{d_{ii}^2} \qquad \text{(for } \mathbf{W} = \mathbf{S}) \tag{2.36}$$

For sufficiently larger L and N_0, the AMSE approaches zero. There is no sytematic error inherent in this method; that is, the AMSE is independent of the input spectrum. The resulting spectrum-estimate $\hat{\psi}$ is unbiased.

Procedure.   In review, the D-method for a singly encoded HT spectrometer is carried out as follows. (Refer to Fig. 2.10, realizing that $\mathbf{W}^{-1} = \mathbf{S}^{-1}$.)

1. Characterize the encoding mask by performing a grating scan or other appropriate procedure, obtaining $\mathbf{T}_o$ and $\mathbf{T}_t$.
2. Compute $\mathbf{D}^{-1} = (\mathbf{T}_t - \mathbf{T}_o)^{-1}$.
3. Perform the intended N spectrometric measurements, obtaining the measurement vector $\boldsymbol{\eta}$.
4. After turning all mask elements "off," take N_0 measurements and average to obtain $\hat{b}$.
5. Subtract $\hat{b}$ from each component of $\boldsymbol{\eta}$ to obtain $\boldsymbol{\eta}'$.
6. Obtain $\hat{\psi}'$ from $\boldsymbol{\eta}'$ via the FHT.
7. Obtain the spectrum-estimate as

$$\hat{\psi} = \mathbf{D}^{-1}\hat{\psi}'$$

Steps 1 and 2 need to be carried out, for a particular spectrometer, only as often as necessary for calibration.

2.4.3. The T-Method

The T-method[37] is a computationally efficient procedure for recovering the optimal unbiased linear spectrum-estimate $\hat{\psi}$ from HT spectrometers having nonideal masks. The development of this method begins in the same manner as that of the D-method. The computation of the optimal estimate $\hat{\psi}$ may be performed in two steps: first, solve Eq. (2.27); then, solve Eq. (2.28). Our goal is to develop an efficient algorithm to accomplish the task.

We saw in the previous section that a preliminary spectrum-estimate $\hat{\psi}'$ can be obtained from the N available measurements via the FHT if $\mathbf{W} = \mathbf{S}$. The task remaining is that of solving Eq. (2.28) efficiently. The

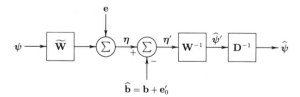

FIGURE 2.10. Block diagram of a multiplexing spectrometer that employs the D-method.

Sherman–Morrison formula[41] provides an elegant vehicle, yielding an expression for $\mathbf{T}^{-1}$ as

$$\mathbf{T}^{-1} = \mathbf{D}^{-1} - \left(\frac{1}{1 + \tau_o^T \mathbf{D}^{-1}\omega}\right)\mathbf{D}^{-1}\omega\tau_o^T\mathbf{D}^{-1} \tag{2.37}$$

where $\mathbf{T}$ is as given in Eq. (2.18).

For $\mathbf{W} = \mathbf{S}$, Eq. (2.37) can be, after some manipulation, rewritten as

$$\mathbf{T}^{-1} = \mathbf{D}^{-1} - q\mathbf{D}^{-1}\mathbf{1}\tau_o^T\mathbf{D}^{-1} \tag{2.38}$$

where

$$q = \frac{2}{(N+1) + 2\sum_{i=1}^{N}(\tau_{oi}/d_{ii})} \tag{2.39}$$

Note that as N becomes large, q approaches zero, which yields

$$\mathbf{T}^{-1} \cong \mathbf{D}^{-1} \qquad (N \text{ large})$$

By substituting Eq. (2.38) into Eq. (2.28) we can obtain the expression

$$\hat{\psi} = \mathbf{D}^{-1}(\hat{\psi}' - q\mathbf{1}\tau_o^T\mathbf{D}^{-1}\hat{\psi}') \tag{2.40}$$

Figure 2.11 shows, in block-diagram form, the implementation as described by Eq. (2.40). We can simplify Eq. (2.40) by introducing a scalar

$$r = \tau_o^T\mathbf{D}^{-1}\hat{\psi}'$$

$$= \sum_{i=1}^{N}\frac{\tau_{oi}\hat{\psi}_i'}{d_{ii}}$$

which is both spectrum- and mask-dependent. Equation (2.40) then becomes

$$\hat{\psi} = \mathbf{D}^{-1}(\hat{\psi}' - (qr)\mathbf{1}) \tag{2.41}$$

The N values that make up the spectrum-estimate given by Eq. (2.41) are given by

$$\hat{\psi}_i = \frac{\hat{\psi}_i' - (qr)}{d_{ii}}, \qquad i = 1, ..., N \tag{2.42}$$

where $\hat{\psi}_i'$ is the ith element of the preliminary estimate obtained via

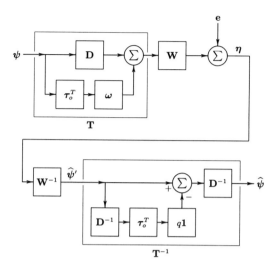

FIGURE 2.11. Block diagram of a multiplexing spectrometer that employs the T-method.

Eq. (2.27), and d_{ii} is the ith diagonal element of $\mathbf{D}$. The product (qr) is computed only once for each run.

It is advantageous to perform certain computations only once for each particular spectrometer, store the results, and use them as needed. For example, $\mathbf{D}$ can be computed ahead of time once the encoding mask has been characterized. $O(3N)$ operations are required to complete such one-time calculations.

For each experimental run, $O(N \log_2 N)$ operations are required in order to compute the preliminary spectrum-estimate $\hat{\psi}'$. Then $O(4N)$ additional operations are required in order to correct for the nonideal transmittances of the mask. Thus, after the initial setup, only $O(N(\log_2 N + 4))$ operations are required in order to obtain the final spectrum-estimate.

The AMSE for the T-method, when $\mathbf{W} = \mathbf{S}$, can be found from Eqs. (2.9) and (2.10) as

$$\epsilon = \frac{\sigma^2}{N} \operatorname{tr}\{\mathbf{T}^{-1}\mathbf{S}^{-1}(\mathbf{S}^{-1})^{\mathrm{T}} (\mathbf{T}^{-1})^{\mathrm{T}}\}$$

An alternative representation for ε is[36]

$$\epsilon = \frac{\sigma^2}{N} \left(\frac{2}{N+1}\right)^2 [(N+1)\operatorname{tr}\{\mathbf{T}^{-1}(\mathbf{T}^{-1})^{\mathrm{T}}\} - \operatorname{tr}\{\mathbf{T}^{-1}\mathbf{J}(\mathbf{T}^{-1})^{\mathrm{T}}\}] \quad (2.43)$$

This expression in turn yields[38] the AMSE for the case of no coaddition (i.e., when $L=1$) as

$$\epsilon = \epsilon(1)$$

$$= \frac{\sigma^2}{N}\left(\frac{2}{N+1}\right)^2 \left\{\left[\sum_{i=1}^{N} d_{ii}^{-2}\right]\left[N + 2q\sum_{j=1}^{N} \tau_{oj} d_{jj}^{-1}\right.\right.$$

$$\left. + q^2\left\{(N+1)\sum_{j=1}^{N} \tau_{oj}^2 d_{jj}^{-2} - \left[\sum_{j=1}^{N} \tau_{oj} d_{jj}^{-1}\right]^2\right\}\right]$$

$$\left. - 2(N+1)q\sum_{j=1}^{N} \tau_{oj} d_{jj}^{-3}\right\} \qquad (2.44)$$

There is no systematic error associated with this method.

For the case of L coadditions, the AMSE is

$$\epsilon(L) = \frac{\epsilon(1)}{L} \qquad (2.45)$$

For large N,

$$\epsilon(L) = \frac{4\sigma^2}{L(N+1)^2}\sum_{i=1}^{N} \frac{1}{d_{ii}^2} \qquad (2.46)$$

Procedure.   A summary of the T-method for a singly encoded HT spectrometer follows. (Refer to Fig. 2.11, noting that $\mathbf{W}^{-1} = \mathbf{S}^{-1}$.)

1. Characterize the encoding mask by performing a grating scan over relevant frequencies, obtaining τ_{ti} and τ_{oi} for $i = 1, ..., N$.
2. Perform the following one-time computations for each spectrometer:

$$d_{ii} = \tau_{ti} - \tau_{oi}, \qquad i = 1, ..., N$$

$$y_i = \frac{\tau_{oi}}{d_{ii}}, \qquad i = 1, ..., N$$

$$q = \frac{1}{((N+1)/2) + \sum_{i=1}^{N} y_i}$$

3. Carry out the intended N spectrometric measurements, obtaining the measurement vector $\boldsymbol{\eta}$.
4. Obtain the initial estimate $\hat{\boldsymbol{\psi}}'$ from $\boldsymbol{\eta}$ via the FHT.

5. Compute $(qr) = q(\sum_{i=1}^{N} y_i \hat{\psi}_i')$.
6. Obtain the spectrum-estimate as

$$\hat{\psi}_i = \frac{\hat{\psi}_i' - (qr)}{d_{ii}}, \qquad i = 1, ..., N$$

Again, steps 1 and 2 need to be performed only once. The remaining steps in the procedure allow for optimal spectrum-recovery, relying on only the N measurements required in step 3.

2.4.4. An Example[38]

In this section we compare the spectrum-recovery abilities of the S-, D-, and T-methods for an example input spectrum. A grating scan, shown in Fig. 2.12, from a prototype electro-optic mask was used in the computer simulation. In the simulation, a known input spectrum was passed through the encoding mask, and then noise was added to simulate the detector noise. Next, each of the three spectrum-recovery methods was applied in conjunction with coaddition. The number of mask elements was $N = 255$. An ideal monochromator was simulated and is used as a point of reference in the comparisons. Recall that an ideal single-slit monochromator provides

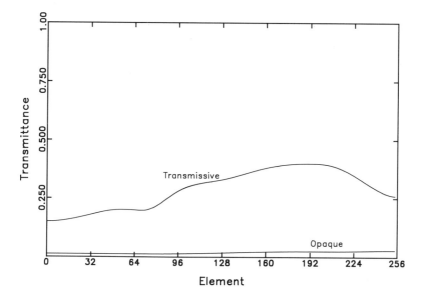

FIGURE 2.12. Grating scan of an example electrooptic mask.

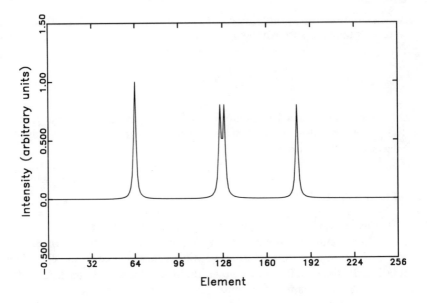

FIGURE 2.13. Input spectrum consisting of Lorentzian peaks.

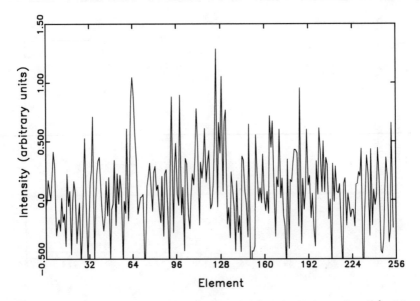

FIGURE 2.14. Input spectrum of Fig. 2.13 corrupted by noise having variance $\sigma^2 = 0.1$.

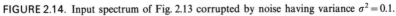

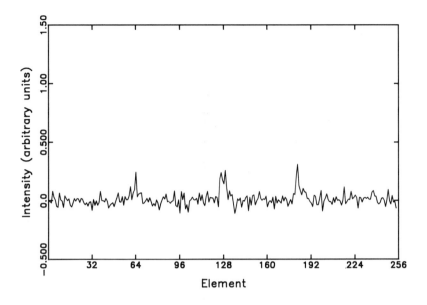

FIGURE 2.15. Spectrum-estimate of Fig. 2.13 obtained via the S-method without coaddition.

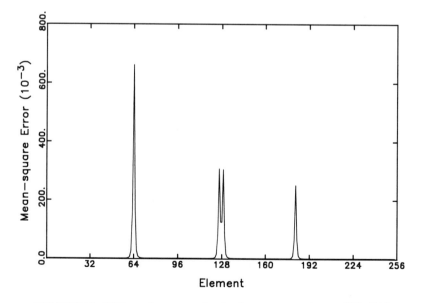

FIGURE 2.16. MSE vs. element number for the spectrum-estimate of Fig. 2.15.

the unbiased spectrum-estimate $\hat{\psi} = \psi + \mathbf{e}$, and has AMSE equal to the variance σ^2 of the noise vector $\mathbf{e}$.

The input spectrum for the simulation is shown in Fig. 2.13. The four peaks are Lorentzian, two of them having been placed to form a doublet. The amplitude of each peak in the doublet is 0.8; the heights of the other two peaks are 1.0 and 0.8. The noise is zero-mean and has variance 0.1. Figure 2.14 depicts the spectrum-estimate that would be obtained via an ideal single-slit monochromator. Figures 2.15–2.20 show both the spectrum-estimates obtained by the three spectrum-recovery techniques reviewed and the resulting MSE curves. Table 2.5 lists the AMSE associated with each of the spectrum-estimates.

Figure 2.15 shows the estimate obtained via the S-method. It should be noted that the peaks have been severely attenuated and the relative magnitudes of the original peaks have been altered drastically. Specifically, the amplitudes of the four peaks at elements 64, 126, 130, and 182 have been changed from 1.0, 0.8, 0.8, and 0.8 to 0.25, 0.25, 0.28, and 0.33, respectively. The above changes are expected, since the difference in transmittance between the transmissive and opaque states of the mask used increases in the range between elements 0 and 182. Figure 2.16 shows the MSE of the estimate in Fig. 2.15. The four peaks are caused by the systematic error. As given in Table 2.5, there is a stochastic error of 0.001556, distributed uniformly over the entire spectral range. That

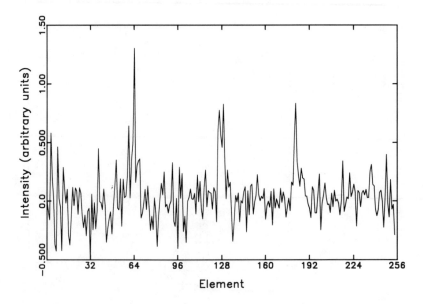

FIGURE 2.17. Spectrum-estimate of Fig. 2.13 obtained via the T-method without coaddition.

component of the MSE cannot be discerned in Fig. 2.16 due to the scale used.

Figure 2.17 depicts the spectrum-estimates obtained via the D-method and the T-method. (Since, when N is large, the spectrum-estimates provided by the D-method are essentially identical to those provided by the T-method, only one plot is shown.) Apparently, the effect of the noise has been reduced. The MSE due to recovery via the T-method is shown in Fig. 2.18. A comparison of Fig. 2.18 with the grating scan of Fig. 2.12 reveals that the MSE is inversely related to the difference between the transmittances of the mask in its transmissive and opaque states. According to Table 2.5, the AMSE of the S-method without coaddition is less than that of the T-method without coaddition. When a large noise variance (in this case $\sigma^2 = 0.1$) and a relatively strong input spectrum are present, the S-method may have a smaller total AMSE than the T-method. However, the estimate obtained via the S-method may, in the vicinity of the spectral peaks, have a much larger MSE than the estimate obtained by the T-method. Compare Figs. 2.16 and 2.18.

The use of coaddition will decrease the AMSE. With only 5 coadditions, the AMSE for the T-method is smaller (0.005769 vs. 0.010235) than that for the S-method. Figure 2.19 depicts the spectrum-estimate obtained via the S-method with 100 coadditions. Note that while the noise has been suppressed dramatically, the amplitudes, both absolute and relative, of the

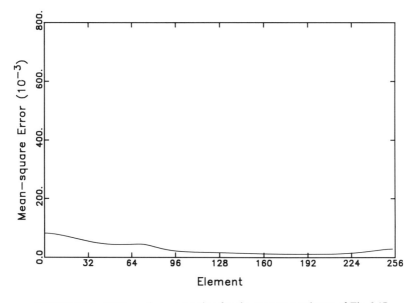

FIGURE 2.18. MSE vs. element number for the spectrum-estimate of Fig. 2.17.

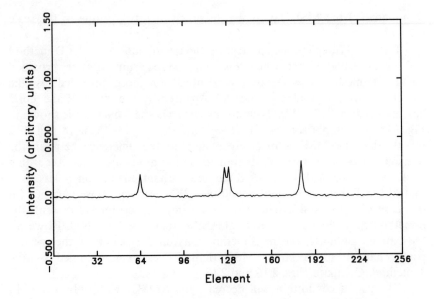

FIGURE 2.19. Spectrum-estimate of Fig. 2.13 obtained via the S-method with 100 coadditions.

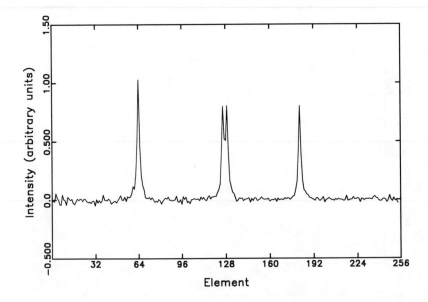

FIGURE 2.20. Spectrum-estimate of Fig. 2.13 obtained via the T-method with 100 coadditions.

TABLE 2.5. Comparison of AMSE for the S-, D-, and T-Methods[a]

| Number of coadditions | Average mean-square error | | | | |
| | S-method | | | | |
	Stochastic	Systematic	Total	D-method	T-method
1	0.001556	0.009924	0.011480	0.034615 ($N_0 = 5$)	0.028845
5	0.000311	0.009924	0.010235	0.011538 ($N_0 = 5$)	0.005769
25	0.000062	0.009924	0.009986	0.002308 ($N_0 = 25$)	0.001154
100	0.000016	0.009924	0.009940	0.000577 ($N_0 = 100$)	0.000288
1000	0.0000016	0.009924	0.009926	0.000058 ($N_0 = 1000$)	0.000029

[a] An input spectrum of four Lorentzian peaks was used. Variance of the detector noise was $\sigma^2 = 0.1$.

peaks are altered significantly from those of the input spectrum. Figure 2.20 shows the spectrum estimated by the T-method with 100 coadditions. The noise has been reduced significantly in this case also, and the amplitudes of the peaks have been recovered with good fidelity.

Finally, as shown in Table 2.5, the AMSE of the T-method is always less than that of the D-method. This will always be the case, regardless of the input spectrum, the characteristics of the mask, and the number of coadditions. Nevertheless, the D-method works essentially as well as the T-method when N is large. Again, the systematic error for the S-method is independent of the number of coadditions used.

2.5. CONCLUSIONS

When the detector noise is the dominant noise in a spectrometric system, an improvement in the AMSE of the spectrum-estimate can be obtained through Hadamard multiplexing. This improvement in AMSE is obtained with the same number of measurements as required by a single-slit spectrometer, thus yielding a multiplex advantage.

In the past, a major obstacle to achieving this multiplex advantage in practice has been the difficulty in precisely positioning a mechanical mask prior to each of the spectrum-measurements performed. However, it is now possible to make stationary masks which are electronically controlled. Unfortunately, the elements of these optical shutters achieve neither perfect transmissivity in the "on" state nor perfect opacity in the "off" state.

Three spectrum-recovery methods have been presented. The S-method has been the method in general use in the past. Unfortunately, it totally ignores any nonidealities in a Hadamard mask, yielding a biased linear estimate of the spectrum. The D-method and the T-method are able to compensate for any nonidealities in the encodement mask, providing unbiased linear estimates. Both of these methods require only a minimal increase in computational effort over that necessary for the S-method. For large N, the AMSE for the D-method asymptotically approaches that for the T-method. The T-method provides an optimal unbiased linear estimate of the true spectrum and, in terms of the overall effort required in measurement and computation, should probably be considered the method of choice.

REFERENCES

1. M. Harwit and N. J. A. Sloane, *Hadamard Transform Optics*, Academic Press, New York (1979).
2. Ref. 1, pp. 1–2.
3. Ref. 1, pp. 2–3.
4. R. M. Hammaker, J. A. Graham, D. C. Tilotta, and W. G. Fateley, "What is Hadamard Transform Spectroscopy?," in: *Vibrational Spectra and Structure* (J. R. Durig, ed.), Vol. 15, pp. 401–485, Elsevier, Amsterdam (1986).
5. D. C. Tilotta, R. M. Hammaker, and W. G. Fateley, "A Visible Near-Infrared Hadamard Transform Spectrometer Based on a Liquid Crystal Spatial Light Modulator Array: A New Approach in Spectrometry," *Appl. Spectrosc.* **41**, 727–734 (1987).
6. A. P. Bohlke, D. Lin-Vien, R. M. Hammaker, and W. G. Fateley, "Hadamard Transform Spectrometry: Application to Biological Systems: A Review," in: *Spectroscopy of Inorganic Bioactivators, Theory and Applications—Chemistry, Physics, Biology and Medicine NATO ASI Series C* (T. Theophanides, ed.), Vol. 280, pp. 159–189, Kluwer, Boston (1989).
7. R. M. Hammaker, W. G. Fateley, and D. C. Tilotta, "Hadamard Transform Spectroscopy: Teaching Old Monochromators New Tricks," *Spectroscopy International* **1** (2), 10–23 (1989).
8. Ref. 1, pp. 6–19.
9. Ref. 1, pp. 200–210.
10. N. Ahmed and K. R. Rao, *Orthogonal Transforms for Digital Signal Processing*, pp. 105–109, Springer-Verlag, Berlin (1975).
11. E. D. Nelson and M. L. Fredman, "Hadamard Spectroscopy," *J. Opt. Soc. Am.* **60**, 1664–1669 (1970).
12. D. C. Tilotta, R. D. Freeman, and W. G. Fateley, "Hadamard Transform Visible Raman Spectrometry," *Appl. Spectrosc.* **41**, 1280–1287 (1987).
13. J. A. Decker, Jr., "Experimental Realization of the Multiplex Advantage with a Hadamard Transform Spectrometer," *Appl. Opt.* **10**, 510–514 (1971).
14. D. C. Tilotta, R. M. Hammaker, and W. G. Fateley, "The Multiplex Advantage in Hadamard Transform Spectrometry Utilizing Solid State Encoding Masks with Uniform Bistable Optical Transmission Defects," *Appl. Optics* **26**, 4285–4292 (1987).
15. P. B. Fellgett, Ph. D. Thesis, University of Cambridge (1951).
16. P. B. Fellgett, "Theory of Multiplex Interferometric Spectrometry," *J. Phys. Radium* **19**, 187–191 (1958).

17. P. B. Fellgett, Aspen Int. Conf. on Fourier Spectr., Aspen, Colorado, 1970 (G. A. Vanasse, A. T. Stair, and D. J. Baker, eds.), AFCRL-71-0019, p. 139 (1971).

18. P. R. Griffiths, *Chemical Infrared Fourier Transform Spectroscopy*, pp. 3–6, Wiley, New York (1975).

19. P. R. Griffiths and J. A. de Haseth, *Fourier Transform Infrared Spectrometry*, pp. v, 78, 220, 248–249, 269, 275–276, 281, 387, 521, Wiley, New York (1986).

20. Aspen Int. Conf. on Fourier Spectr., Aspen, Colorado, 1970 (G. A. Vanasse, A. T. Stair, and D. J. Baker, eds.), AFCRL-71-0019 (1971).

21. P. Jacquinot, 17th Congress du GMAS, Paris, France, p. 25, (1954).

22. P. Jacquinot, "The Luminosity of Spectrometers with Prisms, Gratings or Fabry–Perot Etalons," *J. Opt. Soc. Am.* **44**, 761–765 (1954).

23. P. R. Griffiths and J. A. de Haseth, *Fourier Transform Infrared Spectrometry*, pp. v, 274, 282, 521, Wiley, New York (1986).

24. A. P. Bohlke, J. D. Tate, J. S. White, J. V. Paukstelis, R. M. Hammaker, and W. G. Fateley, "Near-Infrared Hadamard Transform Raman Spectometry," *J. Mol. Struct. (Theochem)* **200**, 471–481 (1989).

25. T. Hirschfeld and B. Chase, "FT-Raman Spectroscopy: Development and Justification," *Appl. Spectrosc.* **40**, 133–137 (1986).

26. D. B. Chase, "Fourier Transform Raman Spectroscopy," *J. Am. Chem. Soc.* **108**, 7485–7488 (1986).

27. C. G. Zimba, V. M. Hallmark, J. D. Swalen, and J. F. Rabolt, "Fourier Transform Raman Spectroscopy of Long-Chain Molecules Containing Strongly Absorbing Chromophores," *Appl. Spectrosc.* **41**, 721–726 (1987).

28. B. Chase, "Fourier Transform Raman Spectroscopy," *Anal. Chem.* **59**, 881A–889A (1987).

29. W. G. Fateley, R. M. Hammaker, D. C. Tilotta, R. Freeman, D. Lin-Vien, A. Bohlke, and J. Linzi, "Hadamard Transform Raman Spectrometry," in: *Proc. XIth International Conference on Raman Spectroscopy* (R. J. H. Clark and D. A. Long, eds.), pp. 941–942, Wiley, Chichester (1988).

30. Ref. 26, Fig. 4.

31. Ref. 12, Fig. 4.

32. Ref. 24, Fig. 1.

33. Ref. 6, Fig. 6.

34. N. A. Vaz, G. W. Smith, and G. P. Montgomery, Jr., "A Light Control Film Composed of Liquid Crystal Droplets Dispersed in an Epoxy Matrix," *Mol. Cryst. Liq. Cryst.* **146**, 17–34 (1987).

35. S. M. Kay, *Modern Spectral Estimation*, pp. 48–49, Prentice-Hall, New York (1988).

36. S. A. Dyer, B. K. Harms, J. B. Park, T. W. Johnson, and R. A. Dyer, "A Fast Spectrum-Recovery Method for Hadamard Transform Spectrometers Having Nonideal Masks," *Appl. Spectrosc.* **43**, 435–440 (1989).

37. T. W. Johnson, J. B. Park, S. A. Dyer, B. K. Harms, and R. A. Dyer, "An Efficient Method for Recovering the Optimal Unbiased Linear Spectrum-Estimate from Hadamard Transform Spectrometers Having Nonideal Masks," *Appl. Spectrosc.* **43**, 746–750 (1989).

38. J. B. Park, T. W. Johnson, S. A. Dyer, B. K. Harms, and R. A. Dyer, "On the Mean-Square Error of Various Spectrum-Recovery Techniques in Hadamard Transform Spectrometry," *Appl. Spectrosc.* **44**, 219–228 (1990).

39. R. A. Dyer, S. A. Dyer, B. K. Harms, T. W. Johnson, and J. B. Park, "Implementation Problems in Hadamard Spectrometry," *IEEE Trans. Instrum. Meas.* **IM-39**, 163–167 (1990).

40. Ref. 1, p. 59.

41. G. H. Golub and C. F. Van Loan, *Matrix Computations*, p. 3, The Johns Hopkins University Press, Baltimore (1983).

Computer-Assisted Methods in Near-Infrared Spectroscopy

3

Tormod Næs and Tomas Isaksson

3.1. INTRODUCTION

Near-infrared (NIR) spectroscopy is now a relatively well established measurement principle[1, 2]. The technique is nondestructive and is based on measurements of reflectance or transmittance of light at different wavelengths in the NIR region (700–2500 nm). From such spectra it is possible to predict chemical parameters such as concentration of fat, water, and protein which are time-consuming to measure by traditional laboratory analyses. In some cases also more complicated measurements, such as the flavor and texture of a food product, can be predicted with relatively high precision.[3]

In Fig. 3.1 a typical NIR spectrum (from a meat sample) is presented. It is seen to be rather smooth with no sharp peaks. This phenomenon is typical for NIR and makes the measurements less sensitive to wavelength drift than in other wavelength regions.

The basic problem in NIR analysis is the calibration of the relation between spectra and the chemical concentrations. First of all the calibration must be multivariate,[4, 5] since it is generally impossible to find selective wavelengths for the different constituents in biological products. In other

Tormod Næs and Tomas Isaksson • Norwegian Food Research Institute, Oslovegen 1, Ås, Norway.

Computer-Enhanced Analytical Spectroscopy, Volume 3, edited by Peter C. Jurs. Plenum Press, New York, 1992.

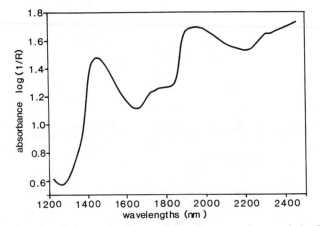

FIGURE 3.1. A typical example of a near-infrared spectrum from analysis of meat.

words, there are a number of interferences in the spectrum making univariate calibration without time-consuming purification of the samples useless.

Traditional (classical) calibration methods based on Beer's law for mixtures[6] usually fail in NIR analysis due to complicated effects, such as differences in light scatter from sample to sample.[1] There are, however, techniques available to correct for such phenomena inside a Beer's law framework,[7] but it is more common to use regression methods with the chemical constituent as the so-called dependent variable and the spectral measurements as independent variables. This is usually called the inverse approach.

Linear regression models are usually satisfactory from a model fitting point of view, but in some cases improvements are obtained by using non-linear regression techniques.[8] Due to extremely high collinearity and often many more wavelengths than samples, it is not always easy to select the calibration method with the best prediction ability. Many different methods have been tried and many of them have proved to be useful, for instance, principal component regression[9, 10] (PCR), partial least-squares (PLS) regression,[11, 12] Fourier regression,[13] and some types of stepwise regression.[2] In this chapter the main attention is paid to principal component regression. It is well understood theoretically, plotting of scores and loadings is easy and can help toward understanding more about structures in the data and to detect outliers.

There are many problems that remain to be solved theoretically and areas for future research will be indicated in the chapter. We will try to show that since theoretical knowledge about certain aspects of NIR analysis is vague, a relatively pragmatic attitude toward the use of statistics may be necessary. This is of special importance in the design of calibration experiments where very little theory is available.[14]

3.2. NOTATION

In this chapter we let y denote the concentration of the chemical constituent for which calibration is conducted. The NIR spectrum for a sample is denoted by x and is assumed to have K wavelengths. The number of samples used for calibration is denoted by I and the number of test (or prediction) samples is denoted by I_p.

The matrix of spectral values for the calibration samples is assumed to be mean centered and denoted by X. The eigenvectors and eigenvalues of X^tX are denoted by $p_1,..., p_k$ and $\lambda_1,..., \lambda_k$ (λ_1 is the largest), respectively. The principal components of X are denoted by $T=\{t_{ik}\}$. Spectra for prediction samples are corrected for the same mean as for the calibration samples.

3.3. SCATTER CORRECTION

Attempts have been made to correct for the more or less irrelevant light scatter variation mentioned in the introduction.[15] The multiplicative scatter correction (MSC)[16] is one of these techniques which has proved

TABLE 3.1. The First 17 Eigenvalues of X^tX
for Uncorrected and Scatter-Corrected
Data[a]

Uncorrected data	Scatter-corrected data
31.49	1.32
4.13×10^{-1}	1.16×10^{-1}
4.46×10^{-2}	5.58×10^{-3}
7.67×10^{-3}	1.27×10^{-3}
1.12×10^{-3}	1.28×10^{-4}
9.07×10^{-4}	4.26×10^{-5}
1.17×10^{-4}	8.27×10^{-6}
2.71×10^{-5}	5.22×10^{-6}
4.49×10^{-6}	3.55×10^{-6}
2.61×10^{-6}	1.78×10^{-6}
1.38×10^{-6}	1.23×10^{-6}
1.22×10^{-6}	6.80×10^{-7}
6.10×10^{-7}	3.39×10^{-7}
5.19×10^{-7}	1.81×10^{-7}
2.27×10^{-7}	6.31×10^{-8}
1.71×10^{-7}	4.44×10^{-8}
6.74×10^{-8}	2.13×10^{-8}

[a] There are 37 spectra and 19 wavelengths in the computations (see Section 3.8). (After scatter-correction, only K-2 ($=17$) eigenvalues are nonzero.)

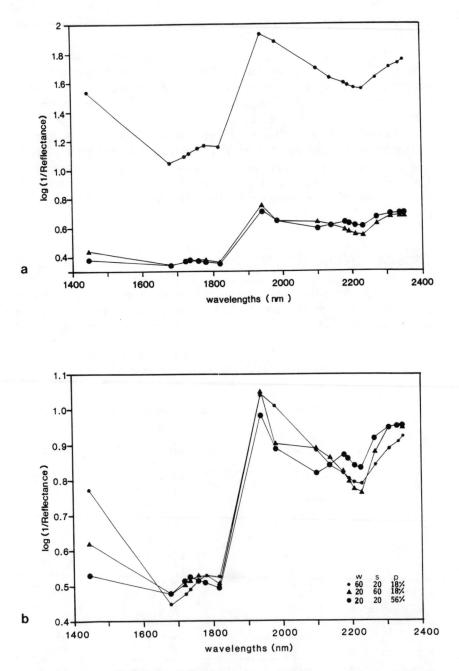

FIGURE 3.2. Three spectra from the calibration set in the example (a) before and (b) after scatter correction (w denotes water, s starch, and p protein in wt%).

useful. This method is motivated partly by spectroscopic theory, which states that the scatter effect is approximately multiplicative, and partly by empirical results. The idea underlying the MSC is to regress the individual spectral values on the average (over samples) spectral values and correct for the slope and intercept differences. Roughly, one can say that the MSC corrects all the samples to the "average scatter level." The MSC transform has been found to represent an advantage from both a prediction and interpretation point of view.[17] In the example below we will therefore concentrate on such scatter-corrected spectra. For other scatter-correction techniques, see Ref. 18 and Ref. 1.

The eigenvalues of $X^t X$ for both the uncorrected and scatter-corrected spectral data in the example below are given in Table 3.1. The scatter correction is seen to have removed a lot of the variation represented in the first eigenvector. Figure 3.2 shows three spectra from the data set in the example before and after scatter correction.

3.4. PRINCIPAL COMPONENT REGRESSION AND NIR ANALYSIS

3.4.1. Model

The statistical model usually applied in NIR spectroscopy is the linear regression model, where y represents the percent of the chemical constituent for which calibration is carried out and x_k are spectral measurements in $\log(1/\text{Reflectance})$ or $\log(1/\text{Transmittance})$ units. This model can be written as

$$y = b_0 + b_1 x_1 + \cdots + b_k x_k + e \tag{3.1}$$

with the usual regression assumptions pertaining to the error terms. The assumptions about error terms should, however, not be taken too rigidly. First, we may have nonlinearities as shown in, e.g., Ref. 17, and second, the concentrations may be selected due to a design strategy (controlled calibration; see Section 3.5).

3.4.2. Selection of Eigenvectors

As discussed above, ordinary least-squares (LS) regression is often of little value in NIR spectroscopy because of collinearity. PCR, as described in, e.g., Refs. 19 and 20, is then a sensible alternative. When a PCR is used in regression, however, one has to decide on a strategy for the selection of eigenvectors or principal components: one can choose from the top or one can use some type of significance test, such as a t-test.[21] In Næs and Martens[22] the first of these strategies is advocated. Roughly speaking, the reason for this is that (at least in sensibly designed experiments) the predictive information is collected in the larger and intermediate eigenvectors, the smaller ones representing mainly noise and irrelevant systematic phenomena.

In more detail, the strategy can be justified by comparing the mean-squared errors (MSE) of prediction for LS regression and PCR with the eigenvectors corresponding to the $(K–A)$ smallest eigenvalues deleted. In terms of principal component scores these can be written as

$$\text{MSE}(\hat{y}_{\text{LS}}) = \sigma^2 \sum_{a=1}^{K} t_a^2/\lambda_a + \sigma^2/I + \sigma^2 \tag{3.2}$$

and

$$\text{MSE}(\hat{y}_{\text{PCR}}) = \sigma^2 \sum_{a=1}^{A} t_a^2/\lambda_a + \left(\sum_{a=A+1}^{K} (t_a/\sqrt{\lambda_a})\theta_a \right)^2 + \sigma^2/I + \sigma^2 \tag{3.3}$$

where θ_a is the regression coefficient for principal component a when this is scaled to have variance equal to one, t_a is used as the new prediction sample's contribution along eigenvector a, and σ^2 is Var(e) in the model given by Eq. (3.1).

It is seen that $t_a/\sqrt{\lambda_a}$ for prediction samples along the deleted eigenvectors and the prediction ability of the same eigenvectors measured by θ_a are the two quantities of importance for the comparison. In other words, both the importance of the eigenvectors in improving the fit and the representativity of the calibration samples for the prediction samples are important for comparing the two predictors.

Computations employing NIR data show that it seems to be typical that θ values corresponding to eigenvectors with high and moderate λ are larger than those for small λ (see the example below, and Refs. 18 and 22). This is mainly due to the fact that for very small λ the eigenvectors are dominated by noise, but it can also be due to irrelevant systematic phenomena with smaller variance than the predictive factors. In addition, it appears to be typical that the value of $t_a/\sqrt{\lambda_a}$ for the eigenvectors with small λ are larger than for eigenvectors with high λ (see Ref. 22 and the example below). The reason for this may be the larger sampling variance for the vectors with small λ.[23]

All this implies that eigenvectors with small λ should be deleted. A compromise between selection strategies could be envisioned, as will be discussed in the example.

3.4.3. Criteria for Selection

How then should the number A of principal components be chosen? Full cross-validation[24] or cross-validation based on the computed principal components of X are both easily interpretable and sensible alternatives. Cross-validation is time-consuming, but PRESS (predicted residual sums of squares) for the other strategy is simple if the residuals divided by one minus the leverage[25] is used. This procedure is often called *leverage correction*, and we have found it good enough in many practical applications (see, e.g., Refs. 10, 17, and 22).

One can, for instance, choose the number of factors with the lowest PRESS value, or the number of factors with a similar PRESS value if this number of factors is smaller. In our experience, the number of factors (A) determined this way is a good estimate even if A is rather large and the variances of some of the principal components are quite small (A may be typically between 5 and 15). We refer to the example below for an illustration.

3.4.4. More Wavelengths Than Samples

An advantage with PCR in NIR analysis is its ability to perform well even with more wavelengths than samples. The reason for this is that the main chemical information is collected in the larger population eigenvectors and to estimate this information not so many samples are needed. For example, Ref. 4 provides an example with 150 wavelengths and only 50 calibration samples, with excellent results in the calibration for protein in wheat.

3.5. DESIGN IN CALIBRATION

Good results with PCR and also with other methods can be obtained if one possesses a "good" calibration set.[2] First, one must ensure that the important information has sufficient variation to be properly separated from the noise. Second, one needs a certain representativity for future samples to avoid too large values of $t/\sqrt{\lambda}$ for prediction samples. Third, one must bear in mind that nonlinearities may occur, and this must be taken care of in the design of the experiment.

In some practical applications of NIR analysis it may be possible to generate artificially a number of samples and use them for calibration. In most cases, however, as, for instance, analysis of meat, it is impossible because meat is a natural product and we cannot control its composition. In such situations it is sensible to try to span out known sources of variation as well as possible by visual inspection of the product or by other prior information. In many cases even this is difficult; one knows too little about the samples to undertake a sensible selection. In such cases one may end up with random sampling as the only alternative. This may lead to good results, especially when the calibration set is moderate or large,[4] but there is always a certain probability that important sample types are not present. Another approach, which can be used in some cases, is to choose samples from measurements of NIR spectra on a large set of samples and submit these samples to chemical analysis and calibration. The idea behind this approach is that NIR analysis is cheap and simple, and spectra are then easy to obtain for a large set of samples. We refer to Refs. 26 and 27 for algorithms of this type.

In this chapter we will consider an example where control of the composition of samples is possible. In Ref. 14 different design strategies for this case are discussed and compared using empirical data. Computation and heuristic arguments based on general principles for design indicated that the following three principles are important:

1. Proper span of the constituents.
2. Presence of all relevant combinations of concentrations.
3. Even spread of design points.

We refer to Ref. 28 for further discussion of point 3, which is here primarily introduced to "smooth out nonlinearities." A factorial design with as many levels as possible for each constituent is suggested as a reasonable strategy.

In the case of "closure" between the absorbing constituents, i.e., the absorbing constituents sum to 100% (which is quite common, e.g., for agricultural products), a real factorial design is impossible to use. A mixture design, such as simplex lattice design,[29] will, however, have properties similar to a factorial design. The simplex lattice structure was used and compared with end-point designs in Ref. 14 and gave good results. The reader is also referred to the example in Section 3.8.

3.6. OUTLIERS IN CALIBRATION

Outliers or abnormalities are sometimes found in applications of NIR spectroscopy. The reasons for this may be several, including a measurement error in y, the instrument being out of order, impurity in the sample, a non-linear relation in a certain subregion, or a sample being from a population other than that anticipated. In calibration, it is of fundamental importance to obtain information about such outliers and their influence on the calibration equation, since outliers can at worst destroy the prediction precision for all future samples. It can, however, also happen that outliers represent valuable information which can improve a calibration (see the example).

A diagnostic procedure for PCR based on two steps was proposed elsewhere.[30] In the first step one searches for influential observations for the eigenvectors used in the PCR. In the second step the observations passing step 1 as "normal" samples are submitted to an ordinary regression diagnostic procedure for the θ values defined in Section 3.3. The motivation for this two-step approach is that it helps to distinguish between different types of abnormality or influence of an observation. For instance, an outlier passing the first step, but which is detected in the second, is one with no significant effect on the principal component structure, but, on the other hand, it has a strange relation between y and x. This observation can be interpreted as one fitting the spectral variation space, but it has either a misprint in y or there is a nonlinear relationship between y and x.

In more detail, the procedure is as follows. In the first step, the diagnostics

$$I_{ij} = -t_{ij} \sum_{\substack{k=1 \\ k \neq j}}^{K} t_{ik}(\lambda_k - \lambda_j)^{-1} p_k I \tag{3.4}$$

proposed in Chritchley[31] for assessing the influence of individual observations (i) on the eigenvectors (j) are computed. The large amount of information within I_{ij} was summarized by considering the lengths of the I_i vectors over the wavelengths; this approach was proposed by Næs.[30]

In the second step, diagnostics denoted by $h_i^{(1)}$, $\hat{f}_i$, and DPC_i are computed from the data for the samples passing the first step. Quantity $h_i^{(1)}$ is defined by

$$h_i^{(1)} = \sum_{a=1}^{A} t_{ia}^2/\lambda_a \tag{3.5}$$

and $\hat{f}_i$ by

$$\hat{f}_i = y_i - \bar{y} - x'\hat{b}_{\mathrm{PCR}} \tag{3.6}$$

while DPC_i is simply Cook's influence measure[32] D_i for the principal component coefficients $\theta_1, ..., \theta_A$ (see Section 3.4). The "leverage" $h_i^{(1)}$ for sample i measures the abnormality of the particular sample within the space of predictive information. In other words, $h_i^{(1)}$ tells how a spectrum is located relative to the main body of the data with respect to the eigenvectors $p_1, ..., p_A$ used in regression. A sample with a large value of $h_i^{(1)}$, for instance, could be one with a chemical composition quite different from the rest. The residuals $\hat{f}$ represent the fit of y to the principal components and give information about, e.g., a printing error in y or a nonlinear relationship between y and x. The DPC_i is analogous to Cook's D_i measure for ordinary LS regression, but is more relevant than D_i when PCR is used as the regression method. Imagine, for instance, that a spectral abnormality is present for an eigenvector which is not used in the PCR calibration. This will affect the ordinary leverage[25] h_i but not the $h_i^{(1)}$ defined as above. Consequently, with similar residuals ($\hat{e}_i \approx \hat{f}_i$), DPC_i is small while D_i is large. In other words, the ith observation has some influence on the least-squares solution, but not on the PCR solution.

Applications of $h^{(1)}$ and $\hat{f}$ for NIR data are treated in Ref. 24.

3.7. OUTLIERS IN PREDICTION

Outlier detection in the calibration step is important in order to obtain precise and reliable predictors. Outliers or abnormalities, like instrument drift and impurity, may, however, be even more frequent in routine use than in a more controlled calibration situation.

For a future sample to be predicted, the only available diagnostic possibility is the spectrum. Below we will advocate the use of

$$h_i^{(1)} = \sum_{a=1}^{A} t_{ia}^2/\lambda_a \tag{3.7}$$

and

$$h_i^{(2)} = \sum_{a=A+1}^{K} t_{ia}^2/\lambda_a \tag{3.8}$$

as useful tools, and point out different reasons for the choice. A spectrum showing up as located far away from the calibration spectra is suspicious, and automatic prediction should usually be avoided for that sample.

First, we consider the formula for predicting the mean-square error of the PCR; it can be expressed as in Eq. (3.3). Prediction ability is seen to decrease with increasing values of $t_a/\sqrt{\lambda_a}$. This means that t values far from 0 are dangerous. It is noteworthy that both t values relative to the eigenvectors used as well as those deleted may be of importance. Consequently, it is not enough to consider $t_a/\sqrt{\lambda_a}$ along the axes used in PCR (i.e., $h_i^{(1)}$). Furthermore, we note that it is not even sufficient to examine $t_a/\sqrt{\lambda_a}$ for the PCR axes and the residuals relative to the eigenvectors used. Imagine, for instance, that an outlier is present in the second eigenvector after truncation. This will show up in the leverage $h_i^{(2)}$ for that space, but not necessarily in the residuals, which are mainly dominated by the first eigenvector after truncation.

The quantities $h_i^{(1)}$ and $h_i^{(2)}$ are very easy to compute as soon as the principal component scores t_{ia} are available. This is certainly a great advantage for routine error warning. In addition, they are easy to interpret as distance measures, and they can be further refined or decomposed to inspect leverage values in the different directions.[30]

Applications of $h_i^{(1)}$ can be found in Refs. 4 and 33.

3.8. EXAMPLE

The scatter-corrected data discussed in Section 3.3 will be used in this illustration. The data set consists of two independently, but identically generated, sets made up of fish meal from cod, starch, and water as described in Fig. 3.3. One of the sets was used for calibration and the other for testing of the prediction equation. The NIR instrument used was a Technicon InfraAlyzer (400) with 19 standard filters possessing prespecified wavelength bands. All the calibrations were performed for protein, which is the dominant constituent of fish meal (about 90%).

As quality criterion of a calibration when tested on new samples we use the root-mean-square error of prediction $(\mathrm{RMSP} = \sqrt{\mathrm{PRESS}/I_\mathrm{p}})$ defined by

$$\mathrm{RMSP} = \sqrt{1/I_\mathrm{p} \sum_{I=1}^{I_\mathrm{p}} (y_i - \hat{y}_i)^2}$$

where I is the number of prediction samples. The measure of model fit is taken as the root-mean-square error of estimation (RMSE) defined by

$$\mathrm{RMSE} = \sqrt{\frac{1}{I - A - 1} \sum_{i=1}^{I} (y_i - \hat{y}_i)}$$

which is just the square root of the ordinary estimate of the error variance in the regression of y on A principal components.

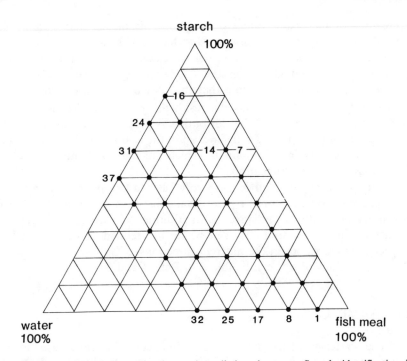

FIGURE 3.3. Design of the calibration and prediction data sets. Sample identification is indicated by numbers 1–37.

The root-mean-square error for the leverage correction is sometimes denoted by RMSEL.

3.8.1. Calibration

The prediction after the calibration is shown in Fig. 3.4 and in Table 3.2. The prediction ability measured by RMSP is seen to decrease to 1.27 for $A = 8$ before it increases to about 2.50. In the leverage correction the lowest value was obtained at $A = 10$. It is seen from the predictions that this would have been a relatively good choice. We note the ability of eigenvectors with eigenvalues down to a very low percent of the largest (Table 3.1) to give predictive improvement.

The $\hat{\theta}$ values in the computation show a clear tendency to be very small for large values of A. In Fig. 3.5 the horizontal line corresponds to a t-test at the 5% level. The average leverage values for the samples (except samples 36 and 37, which have extremely large h_i, see below) are given in Fig. 3.6. There is a clear tendency to increase as A increases. These findings support the arguments in Section 3.3 for selection strategy.

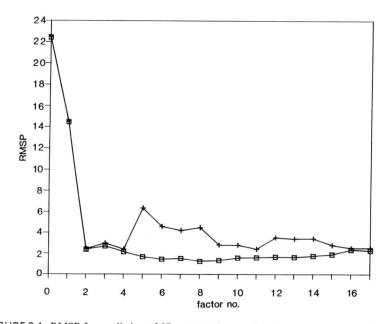

FIGURE 3.4. RMSP for prediction of 37 test samples as a function of the number of factors. Boxes ($\square$) indicate the results from calibration on 37 samples and crosses ($+$) the results from calibration on 35 samples.

The value of $\hat{\theta}_a$ for $a = 7$ is seen to be nonsignificant. We tested the prediction ability of the PCR predictor with eigenvector 7 deleted. Compared to the results obtained with the eigenvector present, this did not yield any improvement.

We could, of course, stop the analysis here with rather satisfactory prediction results, but with limited knowledge about the structure of the different data points and about any possible potential of improvement. We decided to proceed through a diagnostic procedure like that described in Section 3.6.

3.8.2. Diagnostics for the Eigenvectors

First, a study of the leverages as the number of eigenvectors increased revealed that two of the calibration samples were quite different from the rest, namely, sample 36 and sample 37. Already at $A = 6$ the leverage is about 0.75 for both samples.

Compared to the average 0.19 this is quite high. In the next step we computed the Chritchley diagnostics (see below) for the eigenvectors. We decided to pay the main attention to eigenvectors with the largest eigen-

TABLE 3.2. Calibration, Leverage Correction, and
Prediction Results with 37 Samples in the
Calibration Set and 37 Samples in the Prediction Set

Principal components	Fitting (RMSE)	Leverage correction (RMSEL)	Prediction (RMSP)
0	22.76	23.07	22.40
1	14.80	15.11	14.50
2	2.48	2.65	2.40
3	2.33	2.51	2.71
4	2.05	2.21	2.18
5	2.00	2.22	1.69
6	1.46	1.84	1.45
7	1.48	1.88	1.53
8	1.46	1.89	1.27
9	1.19	2.04	1.35
10	0.86	1.11	1.59
11	0.82	1.22	1.63
12	0.83	1.22	1.70
13	0.72	1.28	1.67
14	0.73	1.61	1.79
15	0.70	1.79	1.93
16	0.69	1.43	2.37
17	0.69	1.30	2.29

value, since observations which are influential for these will change the eigenvectors in such a way that diagnostic analyses for eigenvectors with a small eigenvalue do not make much sense. The results from the first seven eigenvectors are reported in Table 3.3. The same two samples are seen to have the most dominant influence on these eigenvectors. Some other samples also showed differences from the rest, but not as extreme as the other two. When relating this influence analysis to the triangle in Fig. 3.2, we see that the most influential cases correspond to observations near the corners. In other words, the samples with some type of extreme position represent a special problem in the sense that they do not fit well into the eigenvector structure estimated by the rest of the samples.

What is the effect of this on prediction? We decided to delete the two most extreme samples in Table 3.3 and recalibrate using the remaining data. The results of this are reported in Table 3.4 and Fig. 3.4. The RMSE is at the same level as above, so there is no indication of a better linear fit with the two samples deleted. Looking at RMSP, however, we see that the values are not as low as above.

In order to find the reason for this, we computed prediction residuals

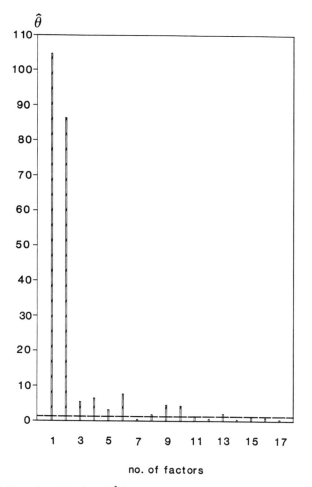

FIGURE 3.5. Bars show the size of $\hat{\theta}$ for $a = 1,..., 19$. The horizontal line corresponds to a t-test at the 5% level.

and in Fig. 3.7 they are plotted for both calibration sets (with 35 and 37 samples in the calibration). It is seen that the residuals for samples 36 and 37 are much larger when the corresponding samples are deleted from the calibration. This clearly shows that these two samples represent dimensions or directions in the spectral space which can be important and which are not well represented in the rest of the samples. We note also from Fig. 3.8 that in both cases the same two samples showed up with larger "prediction leverage" than the rest, which indicates an abnormality of some type. This means that in both cases one should exercise care with automatic prediction of the two samples.

TABLE 3.3. The Length of l_i (Defined in Section 3.6) for Eigenvectors 1, ..., 7 when 37 Samples Are Used in the Calibration[a]

	1	2	3	4	5	6	7
1	—	—	1.4	1.6	—	—	—
2	—	—	—	—	—	—	—
3	—	—	—	—	—	—	—
4	—	—	—	—	—	—	2.5
5	—	—	—	—	—	—	1.9
6	—	—	—	—	—	—	—
7	—	1.3	2.0	1.7	—	—	4.2
8	—	—	—	—	—	—	2.2
9	—	—	—	—	—	—	1.6
10	—	—	—	—	—	—	1.6
11	—	—	—	—	—	—	4.0
12	—	—	—	—	—	—	3.3
13	—	—	—	—	—	—	1.4
14	—	—	—	—	—	—	2.2
15	—	—	—	—	—	—	—
16	—	—	—	—	—	1.3	3.0
17	—	—	—	—	—	—	1.7
18	—	—	—	—	—	—	—
19	—	—	—	—	—	—	—
20	—	—	—	—	—	—	—
21	—	—	—	—	—	—	—
22	—	—	—	—	—	—	—
23	—	—	—	—	—	—	—
24	—	—	—	—	—	1.5	2.6
25	—	—	—	—	—	1.6	2.8
26	—	—	—	—	—	—	1.8
27	—	—	—	—	—	—	—
28	—	—	—	—	—	—	—
29	—	—	—	—	—	—	1.1
30	—	—	—	—	1.7	1.7	—
31	—	—	—	—	2.1	2.9	6.1
32	—	—	—	—	—	—	1.6
33	—	—	—	—	—	—	—
34	—	—	—	—	—	—	2.9
35	—	—	—	—	—	—	3.4
36	—	—	2.4	3.7	5.4	4.3	—
37	—	—	—	—	7.8	9.3	8.0

[a] Only numbers larger than 1 are presented.

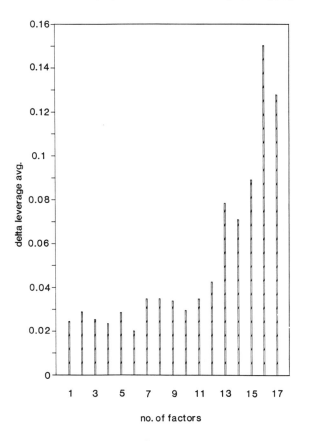

FIGURE 3.6. Bars show the average of the t_a^2/λ_a for the prediction samples (with samples 36 and 37 deleted).

What then about the main body of prediction samples with the two extreme prediction samples deleted? The average prediction results for both calibration sets are presented in Fig. 3.9 and Table 3.5. The results are similar to each other and the best results are obtained at about $A = 12$. The average results in this case are substantially better than above and in fact very good compared to the high variation in the chemical compositions. The conclusion is that for the main body of samples there is no effect if we delete the two calibration samples. We note also that the effect of overfitting is very small in both cases.

Without the additional information in the prediction set, one would probably be careful with predictors influenced by two samples as extreme as 36 and 37 in the calibration, and one would end up with the predictor based on 35 calibration samples.

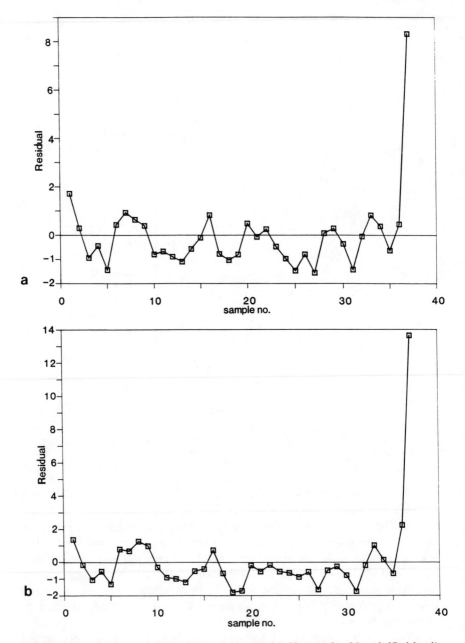

FIGURE 3.7. Prediction residuals when (a) 37 and (b) 35 (samples 36 and 37 deleted) calibration samples are used. In (a) the number of factors, A, is 10 and in (b) the number of factors is 11, which both correspond to the number which would have been selected from leverage correction.

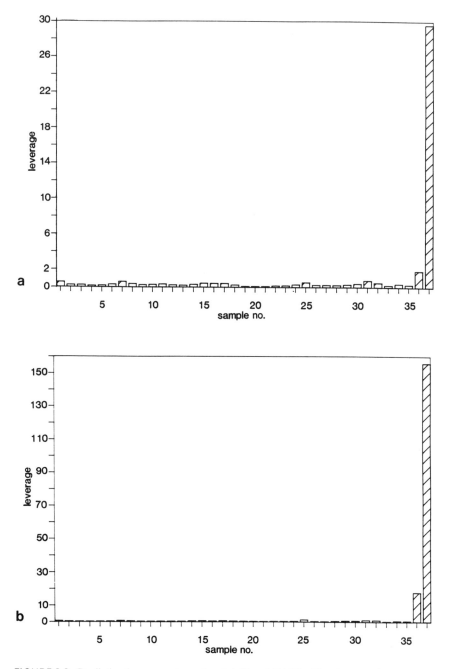

FIGURE 3.8. Prediction leverage values when (a) 37 and (b) 35 calibration samples are used. In (a) the number of factors, A, is 10 and in (b) the number of factors is 11 (see Fig. 3.7).

TABLE 3.4. Calibration, Leverage Correction, and
Prediction Results with 35 Samples in the
Calibration Set and 37 in the Prediction Set

Principal components	Fitting (RMSE)	Leverage correction (RMSEL)	Prediction (RMSP)
0	22.13	22.46	22.50
1	15.20	15.53	14.52
2	2.30	2.46	2.45
3	1.96	2.11	3.00
4	1.68	1.86	2.40
5	1.48	1.63	6.38
6	1.39	1.58	4.59
7	1.41	1.66	4.22
8	1.14	1.94	4.50
9	1.00	1.82	2.83
10	1.06	1.42	2.79
11	0.86	1.08	2.44
12	0.83	1.11	3.55
13	0.71	1.00	3.42
14	0.72	1.08	3.44
15	0.69	1.04	2.81
16	0.70	1.19	2.52
17	0.72	1.10	2.55

We repeated the influence analysis on the 35 remaining samples and observed that some of the samples still had some influence on the eigenvector determination, though not as extreme as above. The most influential cases were collected near the same corner as samples 36 and 37. The influence of several samples on the relevant eigenvectors makes it difficult to perform step 2 in the diagnostic procedure without deleting too large a region of the population, so we decided to terminate the diagnostic procedure here. The prediction results so far are also good enough to be valuable in practice.

3.8.3. Prediction Ability and Leverage

In order to illustrate the idea in Section 3.7, a more detailed study of the relation between "prediction leverages" t_{ia}^2/λ_a and residuals for the first calibration is given in Fig. 3.10. (Samples no. 36 and 37 with the most extreme leverages are deleted so as not to overshadow the information in the rest.) There is good correspondence between the sizes of prediction leverage and prediction error for both $A = 10$ and $A = 17$ (the residuals were computed with $A = 10$ in both cases). It should be noted in particular

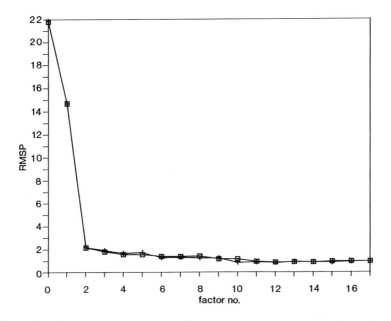

FIGURE 3.9. RMSP computed from 35 prediction samples. Boxes ($\square$) show the results from calibration with 37 samples and crosses ($+$) show the results from calibration with 35 samples.

that if we take into account both $h_i^{(1)}$ and $h_i^{(2)}$, all three samples 1, 25, and 31 would be detected as samples with potentially large residual. This supports the arguments in Section 3.6 advocating the use of both $h_i^{(1)}$ and $h_i^{(2)}$ as diagnostic measures. There are, however, a few exceptions that must be mentioned. First, sample no. 7 has a relatively large h_i value without possessing a large residual. This can, of course, be a coincidence, but it can also be due to precise estimation of θ or a very small θ value in the direction(s) where the sample has a large leverage contribution. We also see that there are a few samples with the opposite effect, namely, large residual and small leverage (samples 5 and 27), which is more difficult to interpret.

3.8.4. Score Plots and Nonlinear Modeling

We saw above how linear regression by PCR yielded satisfactory results. In this section we present a simple investigation of the nonlinearity of relations and the potential for improvement by nonlinear regression. This is an almost neglected area of NIR analysis and research is needed.

In Fig. 3.11 the scores for the first two eigenvectors are plotted, while

TABLE 3.5. RMSP of Prediction with
35 Test Samples when 35 and 37 Calibration
Samples Are Used

Principal components	35 calibration samples (RMSP)	37 calibration samples (RMSP)
0	21.80	21.86
1	14.69	14.66
2	2.15	2.18
3	1.78	1.90
4	1.56	1.67
5	1.56	1.74
6	1.40	1.28
7	1.37	1.30
8	1.41	1.25
9	1.17	1.24
10	1.15	0.84
11	0.93	0.87
12	0.86	0.82
13	0.89	0.92
14	0.87	0.90
15	0.93	0.85
16	0.95	0.92
17	0.95	0.96

the scores for the first three eigenvectors are plotted in Fig. 3.12. The different samples are well separated in both plots and the plot in the first two dimensions resembles the triangle in Fig. 3.3. The relation between the chemical components and the scores in the first three dimensions is, however, obviously nonlinear. The clear and relatively simple patterns in the figure also indicate that the relation between y and the principal components is only moderately complex. This indicates that one could obtain good predictions even for three dimensions if we could find a good transformation or model to use. This would be an advantage, both because a simple plot of the data is advantageous and also because inspection of scores reveal that they are quite stable for similar samples in the first three dimensions.

We tested a simple second-degree polynomial in the scores t_1, t_2, t_3 with interaction terms. The results of this calibration gave a RMSE equal to 1.02 and a RMSP equal to 0.92. The difference between the average estimated and predicted residuals is small, which indicates a valid regression equation. It is also important to note that RMSP is at the same level as the results obtained with PCR after 10 factors with samples 36 and 37 deleted from the prediction. In other words, the problem with samples 36

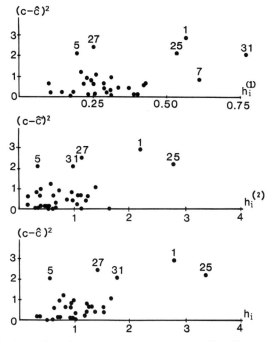

FIGURE 3.10. Squared prediction error at $A = 10$ versus $h_i^{(1)}$, $h_i^{(2)}$, and $h_i = h_i^{(1)} + h_i^{(2)}$.

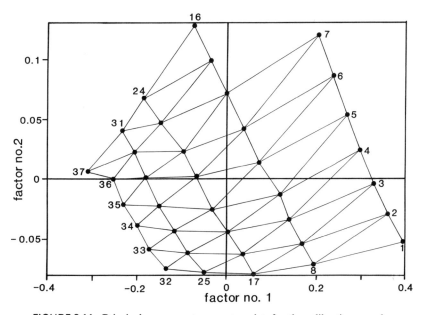

FIGURE 3.11. Principal component scores t_1 and t_2 for the calibration samples.

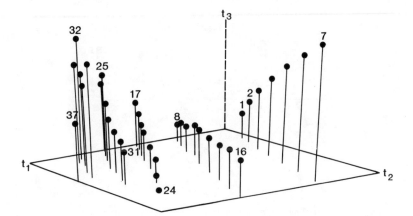

FIGURE 3.12. Principal component scores t_1, t_2, and t_3 for the calibration samples. The t_1-axis extends from -0.31 to 0.40, the t_2-axis from -0.078 to 0.13, and the t_3-axis from -0.02 to 0.03.

and 37 is eliminated by using only 3 factors, indicating the better stability of the scores in the first three dimensions. This is verified by plotting the prediction scores on top of the calibration scores.

The nonlinearity problem is not studied further here, but the results obtained are a clear indication of the possibilities of improvement along this track. (See Ref. 17 for similar results based on projection pursuit techniques, and Ref. 8 for results based on local linear regression.)

ACKNOWLEDGMENTS

We would like to thank Liv Bakke for skillful text-editing and Bjørg N. Nilsen for skillful graphic work.

REFERENCES

1. P. C. Williams and K. H. Norris, eds., *Near-Infrared Technology in the Agricultural and Food Industries*, American Cereal Association, St. Paul, Minnesota (1987).
2. B. G. Osborne and T. Fearn, *Near-Infrared Spectroscopy in Food Analysis*, Longman, Essex (1986).
3. M. Martens and H. Martens, "Near-Infrared Reflectance Determination of Sensory Quality of Peas," *Appl. Spectrosc.* **40**, 303–310 (1986).
4. H. Martens and T. Næs, "Multivariate Calibration by Data Compression," in: *Near-Infrared Technology in the Agricultural and Food Industries* (P. C. Williams and K. H. Norris, eds.), pp. 57–87, American Cereal Association, St. Paul, Minnesota (1987).

5. H. Martens and T. Næs, *Multivariate Calibration*, Wiley, Chichester (1989).
6. E. R. Malinowski and D. G. Howery, *Factor Analysis in Chemistry*, Wiley, New York (1980).
7. T. Næs, "Multivariate Calibration by Covariance Adjustment," *Biom. Journal* **28**, 99–107 (1986).
8. T. Næs, T. Isaksson, and B. Kowalski, "Locally Weighted Regression and Scatter Correction for Near Infrared Reflectance Data," *Anal. Chem.* **62**, 664–673 (1990).
9. I. A. Cowe, J. W. McNicol, and D. C. Guthbertson, "A Designed Experiment for the Examination of Techniques Used in the Analysis of Near Infrared Spectra. Part 2," *Analyst* **110**, 1233–1240 (1985).
10. T. Isaksson and T. Næs, "A Comparative Study of Different Multivariate Calibration Methods on NIR Data," in: *Rapid Analysis in Food Processing and Food Control* (W. Balks, P. Baardseth, R. Norang, and K. Søyland, eds.), pp. 509–513, Norwegian Food Research Institute, Ås (1987).
11. H. Martens and S. Å. Jensen, "Partial Least Squares Regression," in: *Proceedings 7th World Cereal and Bread Congress, Prague, June 1982* (J. Holas and J. Kratochvil, eds.), pp. 607–647, Elsevier, Amsterdam (1983).
12. T. Næs, C. Irgens, and H. Martens, "Comparison of Linear Statistical Methods for Calibration of NIR Instruments," *Appl. Stat.* **35**, 195–206 (1986).
13. W. F. McClure, A. Hamill, F. G. Giesbrecht, and W. W. Weeks, "Fourier Analysis Enhances NIR Diffuse Reflectance Spectroscopy," *Appl. Spectrosc.* **38**, 322–329 (1984).
14. T. Næs and T. Isaksson, "Selection of Samples for Calibration in NIR Spectroscopy," *Appl. Spectrosc.* **43**, 328–335 (1988).
15. G. Birth and H. Hecht, "The Physics of Near-Infrared Reflectance," in: *Near-Infrared Technology in the Agricultural and Food Industries* (P. C. Williams and K. H. Norris, eds.), pp. 1–15, American Cereal Association, St. Paul, Minnesota (1987).
16. P. Geladi, D. McDougel, and H. Martens, "Linearization and Scatter-Correction for Near-Infrared Reflectance Spectra of Meat," *Appl. Spectrosc.* **39**, 491–500 (1985).
17. T. Isaksson and T. Næs, "The Effect of Multiplicative Scatter Corrections (MSC) and Linearity Improvement in NIR Spectroscopy," *Appl. Spectrosc.* **42**, 1273–1284 (1988).
18. A. E. Hoerl, R. W. Kennard, and R. W. Hoerl, "Practical Use of Ridge Regression: A Challenge Met," *Appl. Stat.* **34**, 114–120 (1985).
19. J. Mandel, "Use of the Singular Value Decomposition in Regression Analysis," *Am. Stat.* **36**, 15–24 (1982).
20. R. F. Gunst and R. L. Mason, "Some Considerations in the Evaluation of Alternate Prediction Equations," *Technometrics* **21**, 5–67 (1979).
21. I. Joliffe, *Principal Component Analysis*, Springer-Verlag, New York (1986).
22. T. Næs and H. Martens, "Principal Component Regression in NIR Analysis," *J. Chemometrics* **2**, 155–167 (1988).
23. K. V. Mardia, J. T. Kent, and J. M. Bibby, *Multivariate Analysis*, Academic Press, London (1979).
24. M. Stone, "Cross-validatory Device and Assessment of Structured Prediction," *J. R. Stat. Soc.*, *B* **36**, 111–133 (1974).
25. S. Weisberg, *Applied Linear Regression*, Wiley, New York (1985).
26. T. Næs, "The Design of Calibration in Near Infra-red Reflectance Analysis by Clustering," *J. Chemometrics* **1**, 121–134 (1987).
27. D. E. Honigs, G. M. Hieftje, H. C. Mark, and T. B. Hirschfeld, "Unique Sample Selection in Near Infrared Spectral Subtraction," *Anal. Chem.* **57**, 2299–2303 (1985).
28. P. J. Zemroch, "Cluster Analysis as an Experimental Design Generator, with Application to Gasoline Blending Experiments," *Technometrics* **28**, 39–49 (1986).
29. J. A. Cornell, *Experiments with Mixtures*, Wiley, New York (1981).

30. T. Næs, "Leverage and Influence Measures Related to Principal Component Regression," *Chemolab* **5**, 155–168 (1989).
31. F. Chritchley, "Influence in Principal Component Analysis," *Biometrika* **72**, 627–636 (1985).
32. R. D. Cook and S. Weisberg, *Residuals and Influence in Regression*, Chapman and Hall, New York (1982).
33. T. Næs, T. Isaksson, and H. Kraggerud, "NIR Analysis for Control of Meat Latter Composition," Norwegian Food Research Institute, unpublished (1987).

A Simple-to-Use Method for Interactive Self-Modeling Mixture Analysis

4

Willem Windig

4.1. INTRODUCTION

Despite the use of hyphenated and/or high-resolution analytical instruments, the resulting spectral data often represent mixtures of several components. Furthermore, reference spectra are not always available to resolve the mixture data by techniques such as least-squares or spectral subtraction. For this type of problem, self-modeling mixture analysis techniques have been developed. Generally, the term curve resolution is used for approaches like the one discussed in this paper. But since this technique is not limited to "curve" types of data, such as result from hyphenated instruments, the term self-modeling mixture analysis is used. For an excellent review of factor-analysis-based mixture analysis, see the recent review of Gemperline and Hamilton.[1, 2] For a more geometrical-oriented explanation, see Ref. 3. Most of these self-modeling techniques are based on principal component analysis. Principal component analysis is well suited for this type of problem, because of the noise reduction that can be obtained by the use of the proper number of principal components and because of the ability to

Willem Windig • Chemometrics Laboratory, Eastman Kodak Company, Rochester, New York 14652–3712.

Computer-Enhanced Analytical Spectroscopy, Volume 3, edited by Peter C. Jurs. Plenum Press, New York, 1992.

find pure variables, i.e., variables that have an intensity for only one of the components of the mixtures under consideration.[4]

Although principal component analysis is currently the state-of-the-art approach for self-modeling curve resolution, it is far from a routine laboratory method. A notable exception is Hewlett Packard's Quickres (Infometrix, Inc., 2200 Sixth Avenue, Suite 833, Seattle, Washington 98121),[5] which works with diode-array chromatographic detectors. This approach is, however, limited to resolving two components. There are several reasons for the limited success of principal component analysis as a routine tool.

1. Despite the considerable amount of work done in the area of error analysis,[6-8] there is no established method to determine the proper number of principal components to use.

2. There is great reluctance to use principal component analysis, mainly because principal component analysis does not lend itself easily to the development of user-friendly programs that can be run with the same ease as, for example, library search programs for mass spectral matching.

In order to make self-modeling mixture analysis more accessible for general use, a new method has been developed to resolve mixtures. The new approach is a pure-variable-based method. The pure variable will be determined, however, without the use of principal component analysis. It has been shown before that for curve resolution the pure variables can be determined by simple means.[9] Recently, a new method was described for self-modeling mixture analysis.[10] For the new approach, which will be described here, all the intermediate steps can be presented in the form of spectra and it is possible to direct the search procedure for the pure variables. Knorr and Futrell[4] used data files generated by Ritter[11] in order to evaluate their pure variable method. The same files were used to evaluate the new approach and gave almost exactly the same results.[10] The new approach is simple to use and therefore will be referred to as the SIMPLISMA approach: SIMPLe-to-use Interactive Self-modeling Mixture Analysis. In this chapter, one example will compare the results of the SIMPLISMA approach with results obtained by the principal component analysis based ISMA approach.[12] Furthermore, an example of resolving the data obtained by scanning a polymer laminate (X-ray screen) by FTIR microscopy will be given. Finally, the results of a study of self-modeling mixture analysis of the pyrolysis mass spectra of rubber triblends are presented.

4.2. DATA SETS USED

4.2.1. TMOS Raman Spectra

The first data set results from a study of the formation of silica glasses from solutions of tetra-alkoxysilanes by the sol-gel process, which provides a promising route to low-temperature glass films and monoliths.[13] In order to get a better understanding of the process, the time dependence of the sol-gel hydrolysis and condensation of $Si(OCH_3)_4$ (tetramethyl orthosilicate, abbreviated TMOS) in aqueous methanol is studied. The starting material TMOS has a reported Raman shift at 644 cm^{-1}. The products of this reaction are

1. Hydrolysis products: $Si(OCH_3)_n (OH)_{4-n}$. The Raman shifts for $n = 3, 2, 1$ are 673, 696, and 726 cm^{-1}, respectively.
2. Condensation products: $Si(OSi)_m (OR*)_3$. R* is CH_3 or H. The Raman shifts for $m = 1, 2$ are 608 (shoulder at 586) and 525 cm^{-1}, respectively. These products will be indicated as the first and second condensation products.

This data set was evaluated recently by the ISMA (interactive self-modeling multivariate analysis) program that combined the key-set method with the variance diagram to resolve mixtures.[12] The results obtained by the ISMA approach will be a basis to evaluate the SIMPLISMA approach. This data set will be referred to as the TMOS data set.

4.2.2. X-Ray Screen FTIR Microscopy Spectra

The next data set was obtained by scanning a cross section of an X-Ray screen by FT-IR microscopy. It is a study for the feasibility of using self-modeling mixture analysis to obtain information about the different polymer layers of the X-ray screen without a mechanical separation of the screen. The spectra were recorded using a Digilab IR microscope coupled with an FTS 60 Digilab FT-IR spectrometer. For each spectrum, the observed area was $100\ \mu \times 120\ \mu$ and the samples were moved with a 10-μ step XY motorized stage.

4.2.3. Rubber Triblends Pyrolysis Mass Spectra

The last data set consists of the pyrolysis mass spectra of twelve rubber vulcanizates. The elastomers were styrene-butadiene rubber (SBR), cis-1,4-polybutadiene rubber (BR), and natural rubber (polyisoprene, NR). For a more complete description of this project see Refs. 14 and 15. The

TABLE 4.1. Composition of the Rubber Triblends

Mixture no.	Original wt. %			Mass spectral response		
	SBR	BR	NR	S	BR	NR
1	60	20	20	35	32	33
2	40	40	20	25	40	35
3	20	60	20	13	59	37
4	40	20	40	21	22	57
5	20	40	40	11	29	60
6	20	20	60	9	15	76

composition of these samples is given in Table 4.1. Since the different elastomers had different mass spectral responses (i.e., equal weights of elastomers gave different total ion currents), the weight percents were expressed in mass spectral responses. Furthermore, since the self-modeling mixture analysis will result in resolving the system in Styrene (S) (with SBR as the source), BR (with SBR and BR as sources), and NR rather than the original SBR, BR, and NR, therefore the table gives the expected mass spectral response of S, BR, and NR.

4.3. MATHEMATICAL BACKGROUND OF SIMPLISMA AND ITS APPLICATION TO TMOS RAMAN SPECTROSCOPY DATA

The method will be explained by using a mixture data set presented in Table 4.2. A plot of this data set in a three-dimensional axis system is given in Fig. 4.1. The data points lie in a plane, because the sum of the three concentrations is constant, i.e., one. This causes the loss of one degree of freedom. The plane in which the mixtures lie is limited by a triangle formed by the three pure components.

In practice, a mixture data set will consist of hundreds of variables, such as m/z values for mass spectrometry and wavenumbers of FTIR. Assuming a spectral mixture data set with the same composition as given in Table 4.2, the variables will have contributions from one or more of the components. As a result, every variable will be a positive linear combination of the three component axes. As a consequence, each variable will be bracketed by the three component axes as presented in Fig. 4.1. If each of these variables is presented by a vector the length of which is limited by the triangular plane, the projection of these variables onto the triangle will result in Fig. 4.2, similar to the triangular three-component mixture plots.[3, 12] The position of a vector in this triangle gives a direct measure of the contributions of the three components. All the variables will project

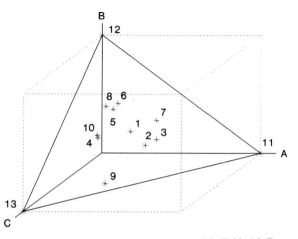

FIGURE 4.1. A three-dimensional plot of the data presented in Table 4.2. Due to the fact that the concentrations of the three components add up to one, all the points representing the mixtures lie in a plane.

within a triangle spanned by the three component axes. A variable that has only contributions from one of the components, a so-called pure variable, will coincide with the component axes. If one does not know the pure variables of a data set and there are reasons to assume the presence of pure variables, it will be clear that the first pure variable can be found by determining the vector with the largest length in the triangular plot.

TABLE 4.2. Composition of the Mixtures in Relative Concentrations

Mixture no.	Relative component concentrations		
	A	B	C
1	0.34	0.34	0.32
2	0.44	0.23	0.33
3	0.48	0.25	0.27
4	0.20	0.35	0.45
5	0.21	0.51	0.28
6	0.22	0.54	0.24
7	0.44	0.37	0.19
8	0.17	0.54	0.29
9	0.33	0.05	0.62
10	0.19	0.37	0.44
11	1.00	0.00	0.00
12	0.00	1.00	0.00
13	0.00	0.00	1.00

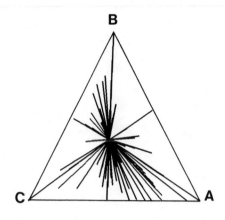

B

C

A

FIGURE 4.2. The projection of vectors representing variables (e.g., wavenumbers) for a spectral data set of a three-component mixture such as presented in Table 4.2 and Fig. 4.1. Pure variables have a relatively large length.

From this plot it is also clear where the other two variables are present. It has to be stressed that the presence of pure components in the data set is not required to obtain this type of result; the only requirement is the presence of pure variables. Triangular plots like this one can be obtained by principal component analysis and form the bases of a visually aided self-modeling curve-resolution method, the variance-diagram technique. This latter technique can be used in combination with the mathematically oriented pure-variable approach to resolve mixtures.[12] Since the lengths of the variable vectors in this triangle are determined by their purity, it is possible to plot a spectrum that represents the purity, using the lengths of these vectors for the intensity axis: a variable with a relatively high intensity will be relatively pure, a variable with a low intensity will have contributions from several components. Furthermore, as mentioned above, the variable with the highest intensity is the first pure variable. The presentation of purity values in the form of a spectrum is more directly related to the original data, lends itself better for a user-friendly approach than the loading type of plots such as shown in Fig. 4.2, and forms the basis for the SIMPLISMA approach. Although the plot shown in Fig. 4.2 is based on principal component analysis, the lengths (purity) of the variable vectors in the triangle can be calculated without the use of principal component analysis, as will be explained below.

The data matrix is represented by D, size $v \times c$, where v is the number of variables and c the number of cases (spectra). The length λ_i of a variable i is given by

$$\lambda_i = \left[(1/c) \sum_{j=1}^{c} (d_{i,j})^2 \right]^{1/2} \tag{4.1}$$

The relation between the length, the mean, and the standard deviation is as follows:

$$\lambda_i^2 = \mu_i^2 + \sigma_i^2 \tag{4.2}$$

where μ_i is the mean of variable i and given by

$$\mu_i = (1/c) \sum_{j=1}^{c} d_{i,j} \tag{4.3}$$

and σ_i is the standard deviation of variable i and given by

$$\sigma_i = \left[(1/c) \sum_{j=1}^{c} (d_{i,j} - \mu_i)^2 \right]^{1/2} \tag{4.4}$$

The vector representation of the relation described by Eq. (4.2) is given in Fig. 4.3. Realizing that the mean of the data set lies in the triangle (see Fig. 4.1), it is possible to rationalize the relation between Fig. 4.1 and Fig. 4.3. The vector labeled μ represents the distance between the origin of that particular variable and the triangular plane; the vector labeled λ represents the vector of that variable with length λ. The vector labeled σ now represents the contribution of that variable in the mixture triangle. The value of σ does not represent the purity as presented in Fig. 4.2 though, since every variable vector will have a different length. A proper way to scale the variable vectors is in such a way that their vectors are

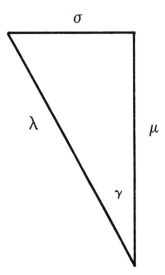

FIGURE 4.3. A vector representation of the relation between the mean (μ), the standard deviation (σ), and the length (λ) of a variable.

limited by the triangular plane as given in Fig. 4.1. Variables scaled in this fashion will project within the triangle, as in Fig. 4.2. As a consequence, the purity of variable i as represented graphically in Fig. 4.3 is defined by

$$p_{i,1} = \sigma_i / \mu_i \qquad (4.5)$$

The second subscript of the purity value p indicates that this is the first purity value. As will be shown below, other purity values will be calculated. The purity value represents $\tan \gamma$, where γ represents the angle between the vector μ_i and λ_i.

At this point, the relation between the purity value as defined in Eq. (4.5) and purity values mentioned in the literature need to be established. The Korr and Futrell[4] method and Malinowski's key-set method[16, 17] define the first pure variable as the one with the lowest projection in the first principal component, after scaling the loadings by their lengths. This is equivalent to $\cos \gamma$ in Fig. 4.3. There is also a relation with a method developed by Vandeginste, in order to determine the purity of a spectrum.[18] In this work, the purest ("simplest") spectrum is defined as the one with the smallest area norm ratio. Calculating the mean and length of rows instead of columns of the matrix results in the area and the norm. As a consequence, the area norm ratio is $(\cos \gamma)^{-1}$. Although the cos and tan are approximately the same for small angles, it will be clear from the knowledge of these geometrical functions that the tan will give more distinct purity values.

The elimination of the effect of the mean vector is also the basis of the calculation of Fig. 4.2. In order to obtain this triangle, the data matrix D was transformed to the matrix $D(\mu)$ in the following way:

$$d(\mu)_{i,j} = (d_{i,j} - \mu_i)/\mu_i \qquad (4.6)$$

The length of the transformed variables equals $p_{i,1}$, as can be derived easily by applying Eq. (4.1) to the transformed elements $d(\mu)_{i,j}$. Principal component analysis on the simulated data set described above results in two principal components, of which the loading plot is the basis of the triangular plot in Fig. 4.2.

Although the method is rationalized above by using a normalized mixture system, it is not necessary to have a normalized system. The principle behind the whole system is that the more independent sources of variance there are, the smaller the standard deviation becomes with respect to the mean: the principle of signal averaging. This is due to the fact that the combined mean of independent sources is the simple sum of the mean values, while the combined variance is the sum of squares of the variance values of the independent sources.

Summarizing, it is possible to calculate the length of the projection of variable vectors onto the triangular plane, as represented graphically in Fig. 4.2 by applying Eq. (4.5). In order to develop and demonstrate the SIMPLISMA method described in this chapter, a data set evaluated before by the principal component analysis based ISMA method will be used. A sample of the spectra is given in Fig. 4.4 Equation (4.5) applied to the data set results in the values plotted in Fig. 4.5a.

The variable with the highest purity appears to be at 700 cm^{-1}. As discussed above, the purity values as given in Fig. 4.5a are based on mean scaled variables. Although this is indeed necessary to determine the purity, it is also convenient to multiply the purity of the variables by their mean values, which facilitates comparison with the original data. Equation (4.5) shows that this will result in the standard deviation spectrum. The spectrum representing the purity values $p_{i,1}$ will be called the first purity spectrum. The spectrum resulting from multiplying $p_{i,1}$ by μ_i will be called the first standard deviation spectrum (see Fig. 4.5b).

The problem now is how to determine the second pure variable. The purity spectrum and the standard deviation spectrum are based on all the components. Because the pure variables are variables that bracket all the other variables in the configuration illustrated by Fig. 4.1, the next pure variable is the one that is most independent of the first pure variable. It is also important to determine the rank of the data matrix, since the rank is determined by the number of components in the data set. The independence of variables and the rank of a matrix can be determined by a determinant-based function. The use of determinants is also the basis of Malinowski's key-set method.[16, 17]

The determinant-based function selected here is called the determinant

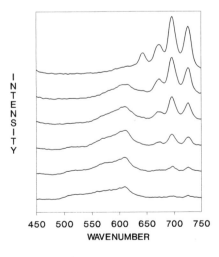

FIGURE 4.4. A sample of the TMOS Raman spectra. The time sequence goes from top to bottom.

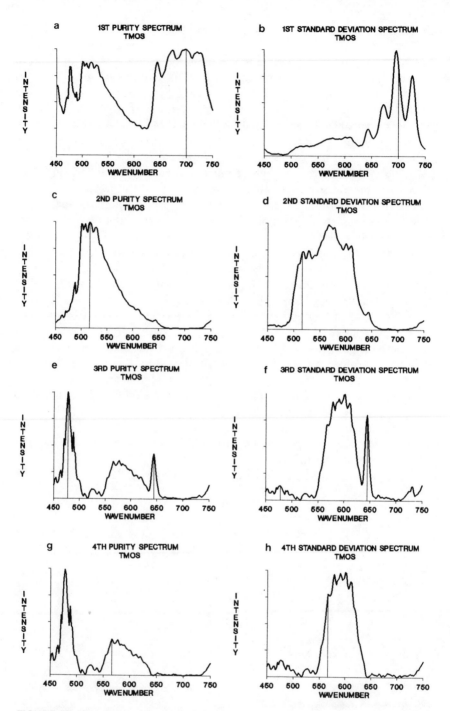

FIGURE 4.5. The purity and standard deviation spectra resulting from the pure variable search by the SIMPLISMA approach. The pure variables are indicated by lines in the spectra.

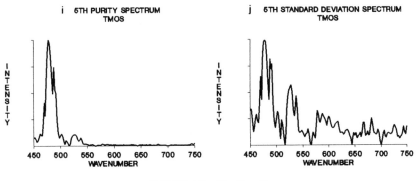

FIGURE 4.5. *(Continued)*

spectrum, of which the elements are calculated as follows. First, the correlation around the origin[19] (COO) matrix needs to be calculated. The COO dispersion matrix has been chosen since it gives all the variables an equal contribution in the calculations. It this is not done, the determinant is not only proportional to the independence of the variables, but also to the length of the variables. In order to calculate the COO matrix, the data matrix first needs to be scaled by the length:

$$d(\lambda)_{i,j} = d_{i,j}/\lambda_i \tag{4.7}$$

The COO matrix now equals

$$C = (1/c)\, D(\lambda)\, D(\lambda)^{\mathrm{T}} \tag{4.8}$$

As a next step, the following determinants are calculated:

$$w_{i,2} = \begin{Vmatrix} c_{i,i} & c_{i,p} \\ c_{p,i} & c_{p,p} \end{Vmatrix} \tag{4.9}$$

Since the matrix C is symmetric and has ones on its diagonal, Eq. (4.9) can be simplified to

$$w_{i,2} = \begin{Vmatrix} 1 & c_{i,p} \\ c_{i,p} & 1 \end{Vmatrix} \tag{4.10}$$

The use of a value 2 for the second subscript of w rather than 1 will be explained below. The values of $w_{i,2}$ will be proportional to the same measure as to which the variables i and p are independent. This can be rationalized by knowing that the determinant calculates the surface area of a parallelogram defined by the row (or column) vectors of the elements of C given in Eqs. (4.9) and (4.10). If variables are highly correlated, the angle between the vector representation of these variables will be small, which will result in a small determinant (surface of parallelogram) defined by these vectors, with a minimum value of 0. Variables that do not have any relation with each other have a maximum value of 1, since the diagonal elements of the matrix in Eqs. (4.9) and (4.10) equal 1 (due to the length scaling). As a consequence, variables highly correlated with the pure variable have a value for $w_{i,1}$ close to 0, while variables which are dissimilar to the pure variable have a high value. The variable with the highest value for the determinant is the next pure variable. The determinant does not express the purity, however. For example, a variable that is shared by the three components as illustrated in Fig. 4.1 will have an intermediate value in the determinant calculation, while it will have a 0 value in the purity calculations. Therefore, the determinant spectrum is used as a weight function for the purity spectrum. Comparative studies showed that this gives better results than the determinant function by itself (not shown). As a consequence, the elements of the purity spectrum now become

$$p_{i,2} = (\sigma_i/\mu_i) \times w_{i,2} \tag{4.11}$$

The resulting spectrum is called the second purity spectrum and is given in Fig. 4.5c. Multiplication by the mean values results in the second standard deviation spectrum (see Fig. 4.5d). The next pure variable is 566 cm^{-1}. In order to find the next (third) pure variable, the determinant function introduced in Eq. (4.9) is now extended to account for the effect of both the first and the second pure variable:

$$w_{i,3} = \begin{Vmatrix} c_{i,i} & c_{i,p} & c_{i,q} \\ c_{p,i} & c_{p,p} & c_{p,q} \\ c_{q,i} & c_{q,p} & c_{q,q} \end{Vmatrix} \tag{4.12}$$

In a manner similar to Eq. (4.11), the third purity spectrum is calculated as follows:

$$p_{i,3} = (\sigma_i/\mu_i) \times w_{i,3} \tag{4.13}$$

or, in a general formulation,

$$p_{i,j} = (\sigma_i/\mu_i) \times w_{i,j} \tag{4.14}$$

where j represents the number of pure variables selected plus one. At this point $w_{i,1}$ needs to be defined. The maximum value for the determinants equals 1, due to the use of variables scaled to unit length. As a consequence, the values for the determinant-based weight function are one before any pure variables have been accounted for:

$$w_{i,1} = 1 \tag{4.15}$$

The maximum in the purity spectrum in Fig. 4.5e is at 478 cm^{-1}. The peak that displays this maximum has an odd shape, however. Furthermore, the standard deviation spectrum (Fig. 4.5f) shows that this wavenumber is in a low-intensity area, which may indicate that this pure variable is caused by noise. A plot of the intensities at 478 cm^{-1} versus time shows an erratic behavior (see Fig. 4.6), while a more or less continuous curve is expected from a time-resolved reaction. In combination with the low intensity it is a strong indication that this is a noise peak. The peak around 478 cm^{-1} is now inactivated for the pure variable search, which can be done interactively with a cursor in the SIMPLISMA program. The next pure variable then is at 644 cm^{-1}. Applying Eq. (4.14) in order to compensate for the first three pure variables results in the fourth purity and standard deviation spectra, given in Figs. 4.5g, h. Ignoring the lower wavenumber range again, the pure variable selected is at 566 cm^{-1}. After eliminating the effect of this pure variable, the fifth purity spectrum and standard deviation spectrum clearly consist of only noise (see Figs. 4.5i, j), which shows that the process is complete. Next to the visual indication that the process of determining pure variables is completed, some functions have been developed. The determinant function will give a value zero, after all the pure variables have been determined. Since the standard deviation spectra are based on the determinant function, their values should also be 0 (or close to 0, due to noise in the data set). The advantage of employing the standard deviation

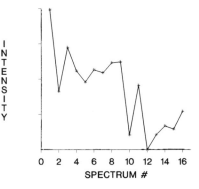

FIGURE 4.6. The plot of the intensities of 478 cm^{-1} in time.

spectrum is that it includes the intensity of the original variables, which avoids a significant contribution of variables in the noise range. The function to determine the rank expresses the value with respect to the first purity spectrum:

$$R_{sj} = 100 \sum_{i=1}^{v} (p_{i,j} \times \mu_i) \Big/ \left(\sum_{i=1}^{v} p_{i,1} \times \mu_i \right) \qquad (4.16)$$

The ratio of the successive values for this function facilitates this process:

$$R_{rj} = R_{sj}/R_{s(j+1)} \qquad (4.17)$$

Since the value of R_{sj} becomes close to 0 after using the proper number of pure variables, the value for this function will be relatively high after determining the proper number of pure variables. Table 4.3 shows that the values of this function also aid in the determination of the rank of the matrix: the fifth value of R_s is relatively low and the fourth value of R_r is high, as a consequence.

It is now possible to resolve the data set, using the pure variable information. Before this is done, a closer look is taken at the noise problem which interfered with the pure variable search described above. The purity value was high in the region around 478 cm^{-1}, while it was relatively low in the standard deviation spectrum (see Figs. 4.5e, f). Since the difference between the purity spectrum and the standard deviation spectrum is a

TABLE 4.3. Values of
Functions to Determine Rank
for TMOS Data

R_s	R_r
100.0000	5.9444
16.8225	9.7213
1.7305	24.0025
0.0721	113.5285
0.0035	0.0006
Noise corrected	
100.0000	9.1018
10.9868	16.8073
0.6737	72.2108
0.0091	97.7223
0.0002	0.0001

multiplication by the mean, it is obvious that the noisy area has relatively low values for the mean. Since the purity value is the ratio of the standard deviation and the mean, it is clear that the high purity value was due to a relatively low mean value. A simple procedure is possible to correct for this. A small value is added to the mean values; this results in the following noise-corrected version of Eq. (4.5):

$$p_{1,j} = \sigma_i/(\mu_i + \alpha) \qquad (4.18)$$

If α has a relatively low value with respect to μ_i (i.e., for high values for μ_i) the effect will be negligible, but for low values of μ_i (in the noise range) the effect is that the purity will be made lower, which is exactly what is needed in order to correct for noise. This simple but effective procedure is similar to the one described by Ozeki et al.[20] Next to this low-intensity correction for the purity value, the same procedure can be applied to the calculations of the determinant, in order to give variables with a low length a lower weight in the calculations. Equation (4.7) then changes into

$$d(\lambda, \beta)_{i,j} = d_{i,j}/(\lambda_i + \beta) \qquad (4.19)$$

The COO matrix is now calculated as given in Eq. (4.8), using these corrected values. The resulting determinants are referred to by $w(\beta)_{i,j}$. As was discussed above, before any pure variables are extracted, the determinant value for each of the variables is 1, which is the maximum value for the determinants. With the correction described in Eq. (4.19), the maximum lengths are different. In order to be consistent, the values for the weight factor for the first purity spectrum need to be calculated as follows:

$$w(\beta)_{i,1} = [\lambda_i/(\lambda_i + \beta)]^2 \qquad (4.20)$$

The reason is that the maximum value for the determinants equals the diagonal elements of the COO matrix. The associated standard deviation spectrum is obtained again by simply multiplying the values by the mean. It appears that a value for α of 5% of the maximum value of μ_i and a value of 5% of the maximum value of λ_i results in purity and standard deviation spectra with the noise virtually eliminated (see Fig. 4.7). The pure variables selected now do not include wavenumbers with intensities in the noise range. The pure variables selected in Fig. 4.7 are close to the ones selected in Fig. 4.5, although the pure variables for the noise-corrected data resulted in a higher similarity with the results obtained by principal component analysis based mixture resolution. The values given by Eqs. (4.16) and (4.17), presented in Table 4.3, appear to be of limited value; it is unclear whether three or four pure variables are present. It is, however, very clear

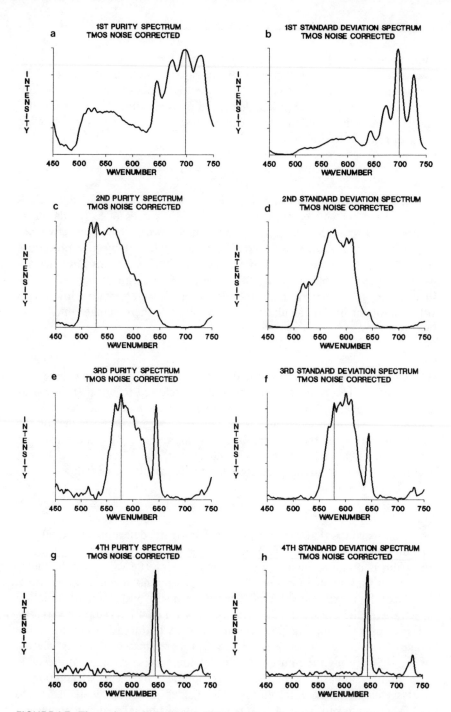

FIGURE 4.7. The purity and standard deviation spectra after correcting for noise. The noise present in Fig. 4.5e is virtually eliminated.

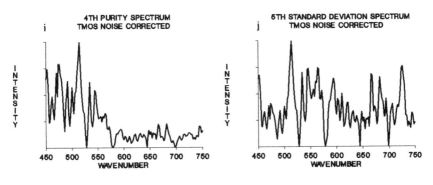

FIGURE 4.7. *(Continued)*

from the fifth purity and standard deviation spectra presented in Fig. 4.7i, h that there are only four pure variables. The problem with the diagnostic values is probably caused by the fact that TMOS, the last component to be resolved in Fig. 4.7, has such a low contribution in the data set, as can be judged from the original spectra in Fig. 4.4. Furthermore, as every spectroscopist knows, a complex pattern such as a spectrum cannot be captured adequately in one single number, which limits the values of functions to determine the rank of a matrix.

With the pure variables available, the data set can be resolved into the pure components and their contributions in the original spectra. The data matrix D can be expressed as a mixture system by the following equation:

$$D^T = CP \tag{4.21}$$

where D^T (size $c \times v$) contains mixture spectra in its rows. The matrix C (size $c \times n$, where n is the number of pure components) contains the relative amount of the pure components in the mixture spectra in its columns and P (size $n \times v$) contains the (unknown) spectra of the pure components in its rows. For the pure variable approach, the intensities of the pure variables in the spectra in D are used in C. As a consequence, D^T and C are known, and P can be resolved by a least-squares method[4]:

$$P = (C^T C)^{-1} C^T D^T \tag{4.22}$$

Assuming that every component results in the same response in the instrument, concentrations can be calculated based on spectra in P that are

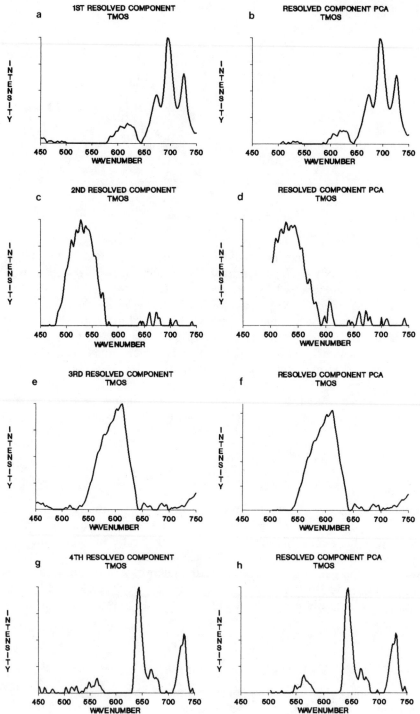

a 1ST RESOLVED COMPONENT
TMOS

b RESOLVED COMPONENT PCA
TMOS

c 2ND RESOLVED COMPONENT
TMOS

d RESOLVED COMPONENT PCA
TMOS

e 3RD RESOLVED COMPONENT
TMOS

f RESOLVED COMPONENT PCA
TMOS

g 4TH RESOLVED COMPONENT
TMOS

h RESOLVED COMPONENT PCA
TMOS

◄

FIGURE 4.8. The spectra in a, c, e, and g are based on the pure variables as extracted by the SIMPLISMA approach (see Fig. 4.7). For comparison the spectra b, d, f, and h, extracted by the principal component analysis based ISMA approach, are given. The results are almost identical.

normalized. This normalization procedure is described by the following equations:

$$f_i = \sum_{j=1}^{v} p_{i,j} \qquad (4.23)$$

The inverse values of f_i are used for diagonal elements of the matrix N, which can then be used to normalize the spectra in P:

$$Q = NP \qquad (4.24)$$

where Q (size $n \times v$) contains the normalized spectra. Although the intensities of the pure variables (after correcting for the normalization procedure described above) can be used for the concentrations, it was decided to obtain the concentrations by a least-squares method, since this results in a noise elimination similar to principal component analysis. Applying the same least-squares approach as described in Eq. (4.22), the least-squares approximation for the concentrations is

$$A = (D^T D)^{-1} D^T Q^T \qquad (4.25)$$

where A (size $c \times n$) now contains the concentrations of the pure components in the mixture spectra.

The spectra and concentration profiles, resulting from the pure variables selected as given in Fig. 4.7, are given in Figs. 4.8 and 4.9, respectively. In order to compare the results obtained before by using the principal component analysis based ISMA approach, these results are also given in Figs. 4.8 and 4.9. The results are almost identical. The first extracted spectrum is associated with the hydrolysis products. Due to a high correlation between the three hydrolysis products, it was not possible to separate them. The pure wavenumber selected by SIMPLISMA was 698 cm^{-1}, which is close to the one reported in the literature, i.e., 696 cm^{-1}. The spectrum in Fig. 4.8b is clearly related to the second condensation product and was calculated on the basis of 528 cm^{-1} as the pure variable, with a reported value of 525 cm^{-1}, which is again very close. The spectrum presented in Fig. 4.8e represents the first condensation product and is based on 578 cm^{-1} as a pure variable, with a reported pure variable

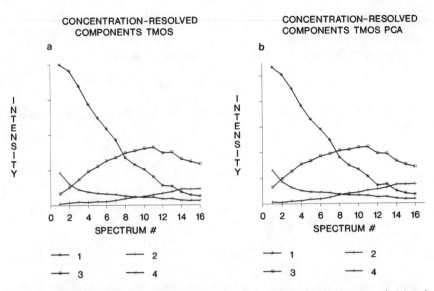

FIGURE 4.9. The concentration profiles as extracted by the SIMPLISMA approach (a) and by the ISMA approach (b). The profile labeled 1 gives the concentrations of the component with the spectrum given in Fig. 4.8a, etc. No significant differences can be observed between the SIMPLISMA and ISMA results.

of 608 cm^{-1} and 586 cm^{-1}. This is a broad peak, which is very likely the cause for the relatively large differences between the value determined by SIMPLISMA and the literature value. The most striking feature, however, is the agreement with the results obtained by principal component analysis (Figs. 4.8b, d, f, h, respectively). Finally, the spectrum in Fig. 4.8d represents the starting material TMOS, with a pure variable at 644 cm^{-1}, which corresponds exactly to the literature value. The concentration profiles as given in Fig. 4.9 are in agreement with expectations and known features.[13]

4.4. OTHER APPLICATIONS OF SIMPLISMA

4.4.1. X-Ray Screen FTIR Microscopy Data

The next example is the X-ray screen data set. For a schematic of the cross section of the X-ray screen, see Fig. 4.10. The expected profile for the components, i.e., PET (poly-ethylene terphthalate), PVA (poly-vinyl acetate), and PVA/PVC (co-polymer of 20% PVA and 80% poly-vinyl chloride), is also given, based on a rectangular smoothing window, in

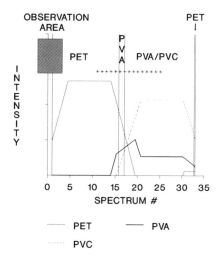

FIGURE 4.10. Schematic representation of the X-ray screen together with an estimate of the concentration profiles of the components, using a rectangular smoothing window with the width of the IR spot, which is indicated by a dashed square. The data points used for the analysis are indicated by asterisks.

order to emulate the effect of the size of the IR spot, which is also given in Fig. 4.10. It is noteworthy that the IR spot is larger than several of the layers. The profile of PVC is given rather than from PVA/PVC, because it is expected that SIMPLISMA separates the PVA as a separate component, since it is also present in another layer. Scanning the edges from the screen gives problems in the spectra, so in order to give a relative high weight to the middle layer only the range indicated in Fig. 4.10 was scanned. A sample of the spectra obtained is given in Fig. 4.11. For many applications, the standard deviation spectra suffice to resolve the spectra, as will be demonstrated by using this data set. Since noise may give

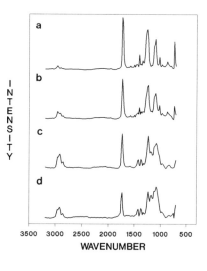

FIGURE 4.11. A sample of the obtained spectra of the X-ray screen.

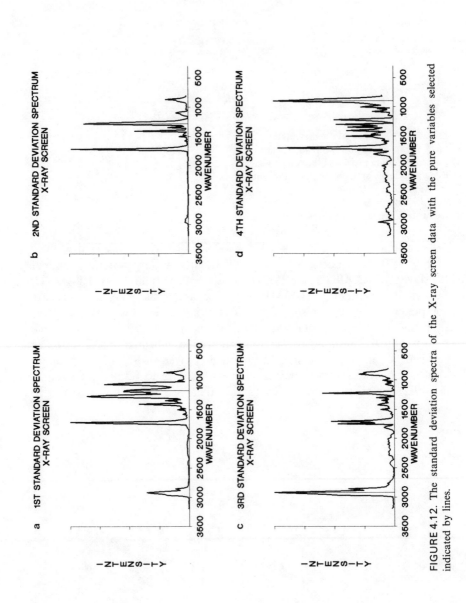

FIGURE 4.12. The standard deviation spectra of the X-ray screen data with the pure variables selected indicated by lines.

problems, the noise level is set at 5% by applying Eqs. (4.18)–(4.20). The pure variable selection is indicated in Fig. 4.12. After selecting three pure variables, the diagnostic values indicate that the process may be finished (see Table 4.4), although there are spectrum-like features in Fig. 4.12d. The pure variable indicated in the fourth standard deviation spectrum appears to be in a range of the spectrum that does not contain any information. Furthermore, resolving the mixtures using this fourth pure variable results in spectra with a significant amount of negative intensities. This knowledge is a strong indication that the rank of the data is 3. The spectra of the pure components and the associated concentrations are given in Figs. 4.13 and 4.14, respectively.

The first extracted spectrum in Fig. 4.13a has similarities with the model spectrum labeled PVA/PVC in Fig. 4.13b. The spectrum in Fig. 4.13b is more complex in nature, because of the phosphor in the screen. Furthermore, since PVA is a separate component in the X-ray screen, it is expected that this first extracted spectrum represents PVA/PVC (+ phosphor) minus PVA. In order to demonstrate this, the other extracted components need to be discussed first. The second resolved component in Fig. 4.13c clearly resembles the model spectrum of PET in Fig. 4.13d. This is not so surprising, since PET is present in a pure form within the resolution of the IR spot in the screen. The last resolved component has similarities with PVA. There is a contribution of the methyl group around 2900 cm^{-1}, which is higher than expected, although PVA is the only source of the methyl group. Possible explanations are interactions, due to the phosphor and/or the fact that PVA/PVC is not really a mixture of PVA and PVC, but a co-polymer. With this extracted PVA spectrum available, it is now possible to test the hypothesis that the first resolved component "lacks" PVA. The sum of the "PVC" spectrum and the PVA spectrum results in Fig. 4.13g, which is almost identical to the model spectrum of PVA/PVC in Fig. 4.13b. These results show that self-modeling mixture analysis can give a very reasonable estimate of the components in a polymer laminate.

The concentration profiles of the extracted components in Fig. 4.14

TABLE 4.4. Values of
Functions to Determine Rank
for X-Ray Screen Data

R_s	R_r
100.0000	4.8269
20.7172	36.8885
0.5616	77.6195
0.0152	

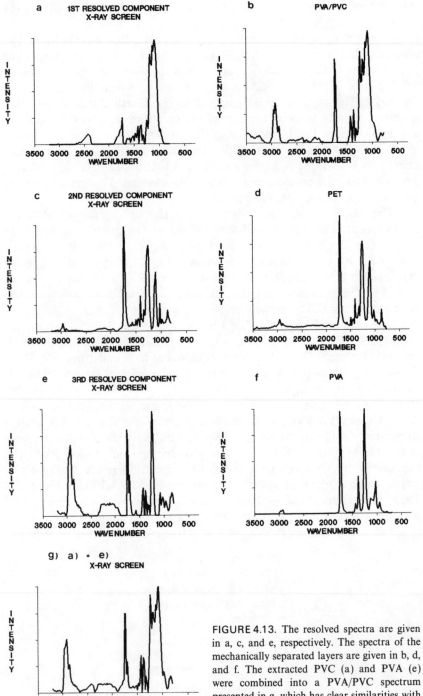

a 1ST RESOLVED COMPONENT
X-RAY SCREEN

b PVA/PVC

c 2ND RESOLVED COMPONENT
X-RAY SCREEN

d PET

e 3RD RESOLVED COMPONENT
X-RAY SCREEN

f PVA

g) a) + e)
X-RAY SCREEN

FIGURE 4.13. The resolved spectra are given in a, c, and e, respectively. The spectra of the mechanically separated layers are given in b, d, and f. The extracted PVC (a) and PVA (e) were combined into a PVA/PVC spectrum presented in g, which has clear similarities with the model spectrum in b.

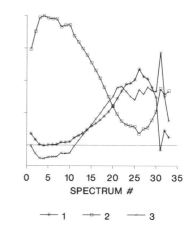

FIGURE 4.14. The concentration profiles of the components, as calculated by the SIMPLISMA approach. The patterns are clearly similar to the estimated profiles in Fig. 4.11. The profile labeled 1 represents the concentration of the component with the spectral pattern in Fig. 4.13a, etc. Especially note the maximum (around spectrum number 20) which is due to the middle PVA layer. The thickness of this layer is clearly below the resolution of the observation area.

include all the spectra. As mentioned above, only part of the spectra was used for the data analysis. It is possible, however, to calculate the least-squares contribution of all the spectra by the proper use of Eq. (4.25). Since the edges of the laminate give problems in the spectra, the first few and the last few spectra have to be viewed with caution. It is obvious, though, that the curves closely match the expected profiles as given in Fig. 4.10. The estimated profiles in Fig. 4.10 are based on the assumption that equal weights of polymers result in equal spectral responses, which is not very likely the case. This is the probable explanation why there are differences between the magnitude of the profiles in Figs. 4.10 and 4.14. It is interesting to note that the PVA profile has a somewhat unexpected behavior in the PVA/PVC layer. The most likely cause of this is the presence of the phosphor particles. The negative intensities in the PVA profile are probably due to negative intensities in the original data set. This is caused by a less than optimal base-line correction procedure.

4.4.2. Rubber Triblends Pyrolysis Mass Spectrometry Data

As a last example the rubber triblends are discussed. The spectra of the pure components are given in Fig. 4.15a–c. Please note that these spectra were not included in the data set (see Table 4.1). Spectra of the mixture samples with the highest amount of the elastomer components (i.e., mixtures 6, 1, and 3, respectively) are given in Fig. 4.15d–f. The standard deviation spectra of this data set are given in Fig. 4.15g–j. The first three pure variables are m/z 136, 104, and 54. The spectra of the pure elastomers (Fig. 4.15a–c) show that these are very likely to present NR, S, and BR, respectively. Although the fourth standard deviation spectrum (Fig. 4.15j) seems to contain information, it appeared that including its pure variable

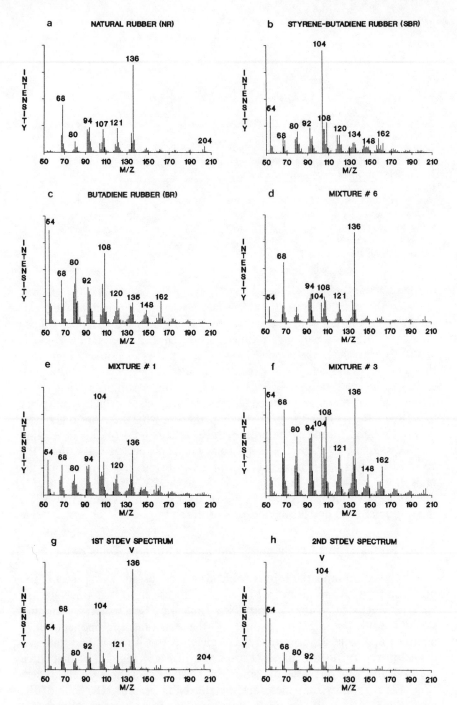

FIGURE 4.15. Pyrolysis mass spectra of the elastomers used for the triblends (a–c) and the spectra with the highest amount of these elastomers (d–f). The standard deviation spectra are

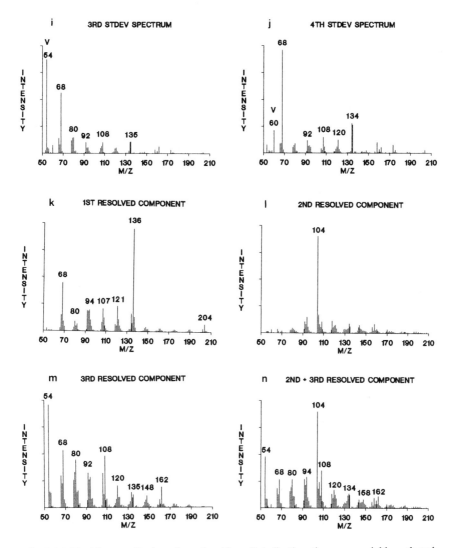

FIGURE 4.15. *(Continued)* given in g–j, with a V indicating the pure variables selected by SIMPLISMA. The resulting resolved spectra are given in k–m (compare with a–c). A combination of l and m results in n, which shows clear similarities with b.

m/z 60 in resolving the mixtures resulted in spectra with negative intensities, from which it was concluded that three pure variables represent the data set properly. The values of Eqs. (4.16) and (4.17) in Table 4.5 appear to be of limited value for this mass spectral data set.

The first resolved component has high similarities with the model spectrum of NR, as can be deduced by comparing Fig. 4.15k and a. The second extracted spectrum (Fig. 4.15l) shows a dominant m/z 104, which represents the styrene monomer. The model spectrum of SBR (Fig. 4.15b) shows a dominant m/z 104 to 0, but also has intensities from the BR contribution, as will be clear from Fig. 4.15c, which shows BR. The third resolved component clearly represents BR (see the reference spectrum in Fig. 4.15c). In order to show that the second resolved spectrum indeed represents the S part of SBR, the second and third resolved spectra were added, which resulted in Fig. 4.15n. Comparison of Fig. 4.15n with the SBR model spectrum (Fig. 4.15b) clearly shows similarities. As a consequence, the conclusion from this study is that the triblends could be resolved successfully into the spectra of the pure components.

The final step is the calculation of the concentrations of the resolved components. Since the samples were analyzed in triplicate, this information was used by substituting the intensities of the individual spectra by their category mean values in matrix C in Eq. (4.22). The least-squares solution will then basically be similar to discriminant analysis. The concentrations resulting from this procedure are plotted versus the values in Table 4.1. The relationships between the expected concentrations and the calculated concentrations are relatively linear, although the values do not coincide with the line that describes the proper relation. Reasons for this are, among others, the fact that the variables are not really pure, which results in offsets as observed in Fig. 4.16. Furthermore, interactions between the products and simplifications and approximations in the calculations between the weight percent and the final mass spectral response are also likely contributors. See Refs. 14 and 15 for more information. The important point is, however, that these results are basically the same as

TABLE 4.5. Values of
Functions to Determine Rank
for Triblend Data

R_s	R_r
100.0000	8.0895
12.3618	10.4419
1.1839	19.1316
0.0619	

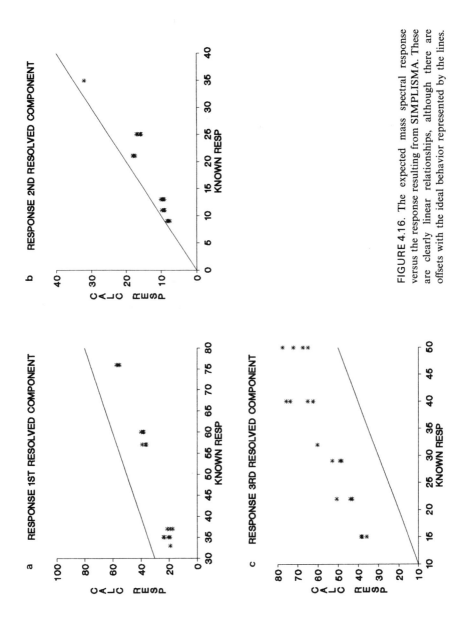

FIGURE 4.16. The expected mass spectral response versus the response resulting from SIMPLISMA. These are clearly linear relationships, although there are offsets with the ideal behavior represented by the lines.

those obtained before by principal component based algorithms, which again shows that the SIMPLISMA approach is a promising tool for self-modeling mixture analysis.

4.5. CONCLUSION

The results of the SIMPLISMA approach shown above and of several other data sets (not shown) are always very comparable with the results of the principal component analysis ISMA approach. At this point, the SIM-PLISMA method did not fail to resolve data sets that were resolved before using the ISMA approach. It is to be expected that, for very noisy data sets, principal component analysis will perform better. A clear advantage of the SIMPLISMA approach is that it displays all the intermediate results in the form of spectra, which clearly facilitates a proper interaction with the operator. As was shown above, an improperly selected pure variable can be detected easily. If desired, the operator can move a cursor to direct the pure variable selection. For example, the first pure variable in Fig. 4.7a is 698 cm^{-1}, which is close to the maximum of a peak in the standard deviation spectrum in Fig. 4.7b. If an operator feels more comfortable with the maximum in the standard deviation spectrum (i.e., 690 cm^{-1}), it can just as well be selected, since the intensity of this wavenumber in the purity spectrum is very close to the maximum. Another possibility is to test the validity of the wavenumbers given as typical for the components: the extracted spectra will show the operator the effect of the selections. Another important point is that the successive steps make intuitive sense (e.g., the noise correction and "eliminating" the effect of selected pure variables) which is not the case for principal component analysis. These results combined make SIMPLISMA promising for routine data analysis in the laboratory, for which no highly skilled operator is needed, as is the case with principal component analysis based method.

Within the industrial environment of Kodak, it is envisioned that SIMPLISMA will be used widely as a routine type of data analysis, while for difficult problems, such as noisy data, components with a very low relative concentration, and data with a lot of interactions, the ISMA approach will have to be used by an experienced analyst. Further developments now in progress are further extensions toward categorized data, which will be comparable to resolving data based on discriminant analysis models and procedures similar to those described by Gemperline[21,22] and Vandeginste[23] to deal with variables that are not pure.

ACKNOWLEDGMENTS

J. L. Lippert is acknowledged for the TMOS project. J. Guilment is acknowledged for the X-ray screen project. R. L. Lattimer (BFGoodrich

Research and Development Center, Brecksville, OH) is acknowledged for his permission to use the rubber triblend data set.

REFERENCES

1. P. J. Gemperline, "Mixture Analysis using Factor Analysis I: Calibration and Quantitation," *J. Chemometrics* **3**, 549–568 (1989).
2. J. C. Hamilton and P. J. Gemperline, "Mixture Analysis using Factor Analysis. II: Self-modeling Curve Resolution," *J. Chemometrics* **4**, 1–13 (1990).
3. W. Windig, "Mixture Analysis of Spectral Data by Multivariate Methods," *Chemometrics and Intelligent Laboratory Systems* **4**, 201–213 (1988).
4. F. J. Knorr and J. H. Futrell, "Separation of Mass Spectra of Mixtures by Factor Analysis," *Anal. Chem.* **51**, 1236–1241 (1979).
5. R. J. Larivee, "Software Review. Quickres," *J. Chemometrics* **3**, 541–542 (1989).
6. E. R. Malinowski, "Determination of the Number of Factors and the Experimental Error in a Data Matrix," *Anal. Chem.* **49**, 162 (1977).
7. E. R. Malinowski, "Theory of the Distribution of Error Eigenvalues Resulting from Principal Component Analysis with Applications to Spectroscopic Data," *J. Chemometrics* **1**, 33–40 (1987).
8. E. R. Malinowski, "Statistical *f*-Tests for Abstract Factor Analysis and Target Testing," *J. Chemometrics* **3**, 49–60 (1988).
9. W. Windig, S. A. Liebman, M. B. Wasserman, and A. P. Snyder, "Fast Self-Modeling Curve Resolution Method for Time Resolved Mass Spectral Data," *Anal. Chem.* **60**, 1503–1509 (1988).
10. W. Windig and J. Guilment, "Interactive Self-Modeling Mixture Analysis," *Anal. Chem.* **63**, 1425–1432 (1991).
11. G. L. Ritter, S. R. Lowry, T. L. Isenhour, and C. L. Wilkins, "Factor Analysis of the Mass Spectra of Mixtures," *Anal. Chem.* **48**, 591–595 (1976).
12. W. Windig, J. L. Lippert, M. J. Robbins, K. R. Kresinske, J. P. Twist, and A. P. Snyder, "Interactive Self-Modeling Multivariate Analysis," *Chemometrics and Intelligent Laboratory Systems* **9**, 7–30 (1990).
13. J. L. Lippert, S. B. Melpolder, and L. M. Kelts, "Raman Spectroscopic Determination of the pH Dependence of Intermediates on Sol-gel Silicate Formation," *J. Non-Cryst. Solids* **104**, 139–147 (1988).
14. R. P. Lattimer, K. M. Schur, W. Windig, and H. L. C. Meuzelaar, "Quantitative Analysis of Rubber Triblends by Pyrolysis Mass Spectrometry," *J. Anal. Appl. Pyrol.* **8**, 95–107 (1985).
15. W. Windig, W. H. McClennen, and H. L. C. Meuzelaar, "Determination of Fractional Concentrations and Exact Component Spectra by Factor Analysis of Pyrolysis Mass Spectra of Mixtures," *Chemometrics and Intelligent Laboratory Systems* **1**, 151–165 (1987).
16. E. R. Malinowski, "Obtaining the Key Set of Typical Vectors by Factor Analysis and Subsequent Isolation of Component Spectra," *Anal. Chim. Acta* **134**, 129–137 (1982).
17. K. J. Schostack and E. R. Malinowski, "Preferred Set Selection by Iterative Key Set Factor Analysis," *Chemometrics and Intelligent Laboratory Systems* **6**, 21–29 (1989).
18. B. Vandeginste, R. Essers, T. Bosman, and G. Kateman, "Three Component Curve Resolution in Liquid Chromatography with Multiwavelength Diode Array Detection," *Anal. Chem.* **57**, 971–985 (1985).
19. E. R. Malinowski and D. G. Howery, *Factor Analysis in Chemistry*, R. E. Krieger Publishing Company, Malabar, Florida (1989).

20. T. Ozeki, H. Kihara, S. Hikime, and S. Ikeda, "Conditions for a Unique Solution in Factor Analysis of Overlapped Raman Spectra of Two Components," *Anal. Sci.* **3**, 285–290 (1987).

21. P. J. Gemperline, "A priori Estimates of the Elution Profiles of the Pure Components of Overlapped Liquid Chromatography Peaks using Target Factor Analysis," *J. Chem. Inf. Comput. Sci.* **34**, 206–212 (1984).

22. P. J. Gemperline, "Target Transformation Factor Analysis with Linear Inequality Constraints Applied to Spectroscopic Chromatographic Data," *Anal. Chem.* **58**, 2656–2663 (1986).

23. B. G. M. Vandeginste, W. Derks, and G. Kateman, "Multicomponent Self Modeling Curve Resolution in High Performance Liquid Chromatography by Iterative Target Transformation Factor Analysis," *Anal. Chim. Acta* **173**, 253–264 (1985).

The Role of NMR Spectra in Computer-Enhanced Structure Elucidation

5

Morton E. Munk

5.1. INTRODUCTION

The elucidation of the structure of an organic compound is of widespread importance in the chemical, biological, and medical sciences. But conventional structure elucidation based on chemical and spectroscopic data can be a complex and time-consuming process, especially in the case of natural products. Such compounds can pose special problems because of their molecular size, skeletal intricacy, and extensive functionalization. However, the specialists that do this kind of work usually assign the correct structure to the unknown.

Thus, although reliability is not a problem in conventional structure elucidation, productivity can be. Since structure elucidation involves the analysis and interpretation of large amounts of data—usually spectroscopic data—the computer, quite naturally, has been the centerpiece of numerous efforts to augment the productivity of the specialist.

As we see it, conventional structure elucidation can be viewed as a process consisting of two separate and distinct stages (Fig. 5.1). The first stage is time-consuming and multistep in nature. It involves collecting

Morton E. Munk • Department of Chemistry, Arizona State University, Tempe, Arizona 85287-1604.

Computer-Enhanced Analytical Spectroscopy, Volume 3, edited by Peter C. Jurs. Plenum Press, New York, 1992.

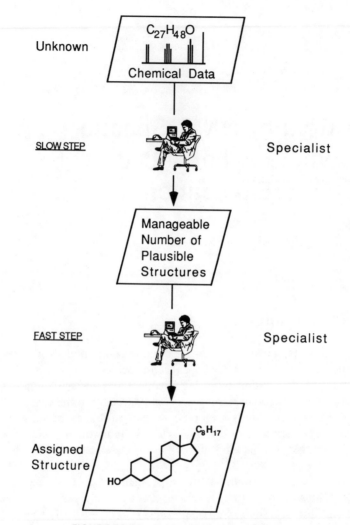

FIGURE 5.1. Conventional structure elucidation.

and interpreting the results of chemical and spectroscopic experiments, most likely running and interpreting additional experiments, and finally expressing the outcome as a *manageable* number of complete molecular structures compatible with all of the data collected. What is a manageable number? It varies with the problem and the specialist, but in general the goal is less than 50. In the second stage, the final assignment comes relatively quickly because specialists are proficient at readily identifying the correct structure of an unknown from among a limited number of alternatives. This step is therefore not the bottleneck; the first step is. If the specialist's

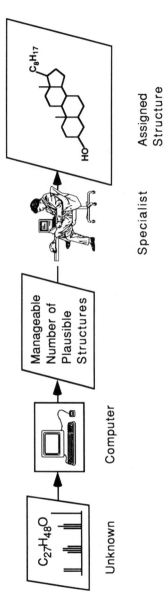

FIGURE 5.2. Goal of the project.

initial involvement could begin with the second stage, much time would be saved and substantially increased productivity achieved.

Thus, the goal of the work described in this paper is the development of computer software capable of directly reducing the collective spectroscopic properties of an unknown to a manageable number of compatible molecular structures (Fig. 5.2). The specialist is still a required player, but he or she is spared the most time-intensive step. The expected outcome is not automated structure elucidation, but rather *computed-enhanced structure elucidation*.

The major target of the software is the compound of complex structure, such as is commonly encountered in natural products. The initial goal is to treat compounds of up to 50 nonhydrogen atoms. If it is met, unknowns of less complexity will of course be readily solved as well.

We note that chemical data have been eliminated as a source of structural information. This is because quite often chemical data are time-consuming to acquire and because it is believed that the collective spectroscopic properties alone, if wisely chosen, can, in many cases, be sufficiently information-rich to narrow the plausible structures to a small number.

The process of structure elucidation includes spectrum interpretation and structure generation as central elements. Thus, the software must possess each of these capabilities. This paper provides an overview of the system under development that highlights some of the ways in which ^{1}H and ^{13}C nuclear magnetic resonance data are utilized in computer-enhanced structure elucidation.

This system is called SESAMI, an acronym for *s*ystematic *e*lucidation of *s*tructure *a*pplying *m*achine *i*ntelligence. SESAMI is a second-generation program and a conceptual departure from its predecessor, CASE.[1]

It must be emphasized that SESAMI is not a finished product and will not be one for some time. It is an evolving, experimental computer program with limited structure elucidation capability at this time. However, the approach appears to offer great promise as illustrated by the examples of problem solving described in Section 5.3.

5.2. PROGRAM SESAMI

5.2.1. Overview

The present organization of SESAMI is shown in Fig. 5.3. INTERPRET is a two-track spectrum interpretation procedure. On one track, the molecular formula and collective spectroscopic properties of the compound of unknown structure are processed to give rise to a set of uniformly-sized,

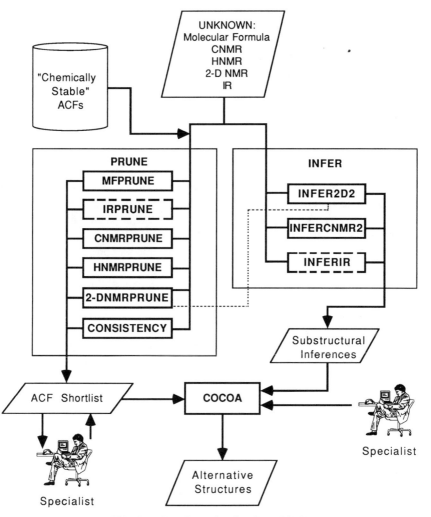

FIGURE 5.3. Information flow in SESAMI.

explicitly-defined basic units of structure that are compatible with the
input. These basic units of structure (the ACF shortlist) can be thought of
as the structural building units for the structure generator, COCOA.[2] On
the other track, the same input leads to substructural inferences in the form
of substructures of any size, and any degree of inexplicitness and overlap.
These inferences serve as the constraints on the structure-generating
process, limiting the number of plausible structures produced. COCOA is
a structure generator that exhaustively constructs all plausible molecular
structures compatible with the dual input.

The structural diversity of compounds found in nature is enormous. In recognition of this, the program must be able to produce the entire range of structures compatible with the observed spectroscopic data, without exception. If the output is to exclude no plausible alternative, the *shortlist* of basic units of structure (ACF shortlist) must exclude no fragment compatible with the spectroscopic data. This can be achieved by taking the set of all possible basic units of structure and deleting from it those that are incompatible with the observed spectroscopic data. That is the role of PRUNE.

This approach to the generation of the shortlist of structural building units imposes two requirements on the nature of the basic unit of structure. First, this fragment must be *small* enough to ensure that the number of all possible basic units of structure is not too large to be efficiently handled by the computer. Second, the fragment must be *large* enough to possess distinctive spectroscopic properties. In practice, these two conflicting requirements cannot be fully met at the same time. A compromise is called for: an atom-centered fragment that has one concentric layer of nearest neighbors (e.g., $=CH-CH_2-O-$). For convenience, this structural unit is referred to as an ACF. The ACF as a unit of structure is not sufficiently large to possess distinctive spectroscopic properties, but early results suggest that it may still serve the intended purpose.

Element groups are the building blocks of ACFs. An element group defines a particular nonhydrogen atom, its attached hydrogen atoms, if any, and the nature of each partial bond by which it can join to other element groups; e.g., all possible element groups for oxygen are hydroxyl oxygen ($-OH$), ether oxygen ($-O-$), and carbonyl oxygen ($=O$). The central element group of the ACF must be valence-satisfied, but at least one first-layer element group must not be, e.g., $(CH_3)_3C-CH=$.

The ACF is a precisely defined substructural unit which can be conveniently and canonically represented for computer manipulation. It can also be correlated to its spectroscopic properties. For initial study, those elements were included that are most commonly encountered in natural products: carbon, trivalent nitrogen, oxygen, divalent sulfur, and each of the monovalent halogens. Important structural features excluded by these elemental limitations can be conveniently added as "super" element groups. The nitro group was included by this mechanism.

The computer-generated exhaustive list of ACFs contained a total of 13,703 fragments. Deleting those that would clearly lead to chemical instability in molecules containing them [e.g., $-CH_2-C(OH)_3$] gave rise to the *exhaustive* list of 5088 *chemically stable ACFs* (Fig. 5.3) from which PRUNE removes those ACFs incompatible with the observed spectroscopic properties of an unknown.

5.2.2. INFER

5.2.2.1. Introduction

INFER is modular in nature and consists of separate routines, the output of each of which is one or more substructures predicted to be present or absent in the unknown (the usual way to express the interpretation of spectroscopic data). No restrictions are placed on the number of substructures inferred, the size of the substructures, the degree of ambiguity in defining the substructure, or the extent to which the substructures derived from the same or different routines may overlap. Alternative substructural interpretations may also be produced.

At the present time, only two substructural inference makers are operational and incorporated into the system: INFER2D and INFERCNMR. Work on an infrared inference maker (INFERIR) is progressing. We note the interactive nature of SESAMI. Structural information known to the specialist from whatever the source can be entered. Thus, the specialist can be viewed as just another substructural inference maker.

5.2.2.2. INFER2D

Two-dimensional NMR spectroscopy is a powerful structural probe and therefore ideally suited as a substructural inference maker for complex compounds.[3] Since there are cases in which the usually acquired two-dimensional NMR data alone are not sufficiently information-rich to narrow the possible structural assignments to a small number, this method complements, rather than replaces, other spectroscopic sources of structural information.

More than any other spectroscopic method, two-dimensional NMR spectroscopy influenced the development of INFER and the design of the new structure generator COCOA. The goal was to utilize *all* of the various structural correlations produced by two-dimensional NMR experiments. Since COCOA currently recognizes only topology and not topography, only through-bond correlations are of value.

The most widely used of such two-dimensional NMR experiments in structure elucidation are $^1H-^1H$ COSY, giving three-bond hydrogen–hydrogen correlations, and long-range $^1H-^{13}C$ COSY, giving long-range (usually two or three bonds) hydrogen–carbon correlations. If these are to be fully and efficiently utilized, $^1H-^{13}C$ COSY, giving one-bond hydrogen–carbon correlations, is also required. The two-dimensional INADEQUATE experiment, giving one-bond carbon–carbon correlations, although very informative, is less widely used because of technical difficulties.

INFER2D is a second-generation program. It is at the same time simpler and more powerful than its predecessor INTERPRET2D.[4] To understand the problems that need to be addressed in interpreting two-dimensional NMR data, it is instructive to first consider INTERPRET2D. The function of this early program as originally envisaged was to reveal carbon–carbon *atom* connectivity, i.e., to produce units of connected carbon atoms present in the unknown. It operates via a two-step process. In the first step of the interpretation process, the program derives carbon–carbon *signal* connectivity, i.e., units of connected carbon *signals*, directly from two-dimensional INADEQUATE data and indirectly from $^1H - ^1H$ COSY/$^1H - ^{13}C$ COSY data. In the absence of molecular symmetry in the unknown, interpretation is complete since then carbon–carbon signal connectivity is the same as carbon–carbon atom connectivity. In the presence of molecular symmetry, algorithms based on group theory perceive symmetries consistent with the entered one-dimensional NMR data (required) and, for each case, convert units of connected carbon signals to one or more discrete substructures. Thus, two or more *different* substructural interpretations are common in the case of unknowns with molecular symmetry.

With this approach a difficult problem is encountered in the reduction of long-range $^1H - ^{13}C$ COSY correlations to a discrete substructure, even in the absence of molecular symmetry. This experiment often produces many separate nonadjacent hydrogen–carbon correlations. In the usual case the number of intervening bonds between the hydrogen and carbon atoms cannot be precisely identified. It is usually two or three bonds, but can be four. Thus, each correlation is ambiguous in the sense that it is consistent with two and possibly three different substructural interpretations, i.e., $H \cdot C \cdot C$, $H \cdot C \cdot A \cdot C$, or $H \cdot C \cdot A \cdot A \cdot C$, where atom A can be any nonhydrogen element. If the goal is to deduce discrete sets of substructures, there is a problem because the set of observed ambiguous correlations will give rise to many different sets of unambiguous correlations. Each such set corresponds to a different set of substructural inferences, i.e., a different interpretation. In the monochaetin problem (Section 5.3) 27 ambiguous hydrogen–carbon correlations were produced, each with two or three intervening bonds. This would give rise to 2^{27} (134 million) different sets of 27 unambiguous correlations. To generate all those different sets is difficult enough. Then the structure generator must cope with that many alternative interpretations of the data.

For similar reasons, chemists, in conventional structure elucidation, often use long-range hydrogen–carbon correlation information *retrospectively* to distinguish between alternative structural assignments for an unknown, rather than *prospectively* to construct compatible structures.

The key to overcoming these obstacles was the recognition that symmetry considerations and multiple interpretations of the data should not be the concern of the interpretation program. Rather, these tasks reside in the domain of structure generation. The output of the interpreter need only be carbon–carbon *signal* connectivity; structure generation should do the rest. COCOA, in contrast to other reported structure generators,[5-10] can process this information directly, because it efficiently perceives symmetry, it uses that information prospectively in building structures, and it uses alternative interpretations of spectroscopic data prospectively.

The current program, INFER2D, only deduces units of connected carbon signals. Specifically, it performs the following simple tasks:

1. It unambiguously determines carbon–carbon signal connectivity from $^1H-^1H$ COSY and $^1H-^{13}C$ COSY data by identifying carbon signals that correspond to carbon atoms bearing coupled vicinal hydrogens.
2. One-bond carbon–carbon correlations (2-D INADEQUATE) are also unambiguously reduced to units of connected carbon signals.
3. Each ambiguous, long-range $^1H-^{13}H$ COSY correlation, for example, those limited to two or three intervening bonds, yields an inference requiring a pair of carbon signals to be either directly connected to one another or separated by one other atom.

Each inference, unambiguous or not, is passed directly to the structure generator COCOA. It should be noted that bond type—single, double, or triple—is not assigned by INFER2D, only connectivity. Additionally, the program does not reveal information about attached hetero atoms.

5.2.2.3. INFERCNMR

INFERCNMR, also a second-generation program, is an interpretive library search routine. Its function is the same as that of its predecessor[11]: to retrieve substructures from the reference compounds of a library of *assigned* ^{13}C NMR spectra that are predicted to be present in an unknown compound. The library *must* contain both spectroscopic and structural information. All library search systems suffer a common problem. Their ultimate performance is database dependent. Substructures not present in the database cannot be retrieved.

Program input consists of the chemical shift of each signal in the spectrum of the unknown and its multiplicity due to one-bond carbon–hydrogen coupling. The molecular formula of the unknown is also needed. The output of the program is one or more explicitly defined substructures predicted to be present in the unknown, and the confidence level of each prediction. Of course, it is possibe that no substructure will be retrieved.

INFERCNMR functions as a subspectrum matching routine. The

premise implicit in such a procedure is that, if a subspectrum of a reference entry matches a subspectrum of an unknown, the substructure assigned to the reference subspectrum is also present in the unknown. Overall structural similarity between the reference compound and the unknown is not necessary in order to retrieve a substructure common to both.

In the initial step of the search strategy, the program retrieves all reference spectra in the library at least four signals of which match signals in the spectrum of the unknown. Currently, matching tolerances up to ± 3.0 ppm are being studied. Matched signals must have the same multiplicity. In a second step, the connection table of each retrieved reference entry is searched to identify those reference subspectra that correspond to a set of at least four connected carbon atoms, that is, to a substructure. In the final step the program estimates the accuracy of each substructure prediction.

A statistical model known as logistic analysis[12] is being studied to develop a measure of prediction accuracy. The results described here are preliminary. Two training sets of 1000 randomly selected reference spectra were taken from a library of 13,000 assigned ^{13}C NMR spectra. For each spectrum in the training sets, predictions were obtained at 30 different matching tolerances, from ± 0.1 ppm to ± 3.0 ppm in increments of 0.1 ppm. Approximately 224,000 substructure predictions were obtained and the validity of each was determined using a substructure matching routine. In addition, values for each of the following eight variables were recorded for each prediction.

1. The molecular size of the reference compound relative to that of the unknown.
2. The root-mean-square chemical shift deviation.
3. The number of residual bonding sites in the predicted substructure.
4. The number of times a given substructure is retrieved from the library.
5. The "size" of the unknown compound.
6. The "size" of the substructure matched.
7. The width of the matching tolerance.
8. The solvent (the same as or different from that used for the reference compound).

Currently, a function based on the first four of these variables [Eqs. (5.1) and (5.2)] has been identified which discriminates between valid and invalid predictions:

$$P_i = \frac{e^X}{1 + e^X} \tag{5.1}$$

where

$$X = \beta_0 + \sum \chi_i \beta_j + \sum \chi_i \chi_k \beta_j + \sum \chi_i^2 \beta_j + \sum \chi_i \chi_k^2 \beta_j \qquad (5.2)$$

with β_0 the intercept, χ a variable, and β_j the weighting coefficient. The latter for each of the four variables [Eq. (5.2)] is derived from the statistical analysis of the data for the 224,000 predictions. In solving for X, the summation is taken over all four variables. In the interaction terms, all possible combinations are taken. (Work on an improved function for X is in progress.)

Equation (5.1) yields P_i (values range from 0 to 1), which is a measure of the probability that a given prediction is correct. By tabulating the actual number of valid and invalid substructure predictions for the set of substructures with P_i values equal to and greater than a given value (e.g., 0.9), a close correspondence between P_i and the actual prediction accuracies observed with each of the two training sets was demonstrated.

Program performance was evaluated using a database of 1300 assigned ^{13}C NMR spectra that were not part of the original reference library. Five sets of 200 randomly selected spectra were generated for testing purposes. The results for each test set are recorded in Table 5.1. Maximum matching tolerance was set at ± 1.5 ppm for this study. We note that at this tolerance the accuracy of the matching routine itself (Column: Matching routine performance, % Valid predictions) varies within a narrow range; 75 to 78%.

The performance summary is intended to reveal the capacity of the probability function to discriminate between valid and invalid predictions.

TABLE 5.1. Evaluation of Performance of INFERCNMR

				Program discrimination					
				% Valid predictions captured			Average # pred./unknown		
Test set	Matching routine performance								
	Total[a]	Valid[b]	% Valid	90%[c]	95%	100%	90%[d]	95%	100%
1	29743	23254	78.2	53.4	20.3	0.6	62	24	0.7
2	21260	16187	76.1	44.9	15.3	3.3	36	12	2.7
3	26554	20342	76.6	46.3	21.9	2.0	47	22	2.1
4	21586	16227	75.2	52.8	30.7	7.5	43	25	6.1
5	23493	18129	77.2	44.7	23.1	3.4	41	21	3.1

[a] Total number of predictions made by the program for each test set of 200 spectra.
[b] Number of valid predictions.
[c] Percent of valid predictions captured at 90% accuracy.
[d] Average number of predictions per unknown at 90% accuracy.

Thus, the program can discriminate between valid and invalid predictions 90% of the time (Column: % Valid predictions captured), but about 50% of the valid information is lost. Table 5.1 shows that the higher the accuracy demanded, the greater the loss of valid information. At 100% prediction reliability, most of the valid information is lost; in this study only 8% or less was retrieved.

In order to examine the amount of information retrieved in another way, the average number of substructure predictions per unknown is also listed. Thus, on the basis of the results with these five test sets, on average, about 21 substructure predictions can be expected for each unknown spectrum entered if the minimum acceptable prediction accuracy is to be 95%. We note that there could be redundancies among these predictions.

5.2.3. PRUNE

PRUNE is responsible for producing the ACF shortlist, the uniformly-sized fragments that are the structural building blocks in structure generation. Like INFER, it is modular in nature.

The starting point for PRUNE is the exhaustive list of chemically stable ACFs. MFPRUNE deletes all ACFs not compatible with the molecular formula of the unknown. IRPRUNE is not fully operational yet. A program is being developed to reliably predict the absence of functionalities easily detected by infrared spectroscopy, for example, the nitro group and the carbonyl group. That information will be used to delete ACFs surviving MFPRUNE.

CNMRPRUNE now takes the list of ACFs surviving MFPRUNE and deletes those carbon-centered ACFs not compatible with the ^{13}C NMR spectrum of the unknown. The routine uses a database containing allowed ^{13}C NMR chemical shift ranges and signal multiplicities for the central carbon atom of all carbon-centered ACFs. The surviving carbon-centered ACFs are organized in a specific manner. The program compiles a list of plausible carbon-centered ACFs for each signal in the ^{13}C NMR spectrum of the unknown. An ACF is assigned to an observed chemical shift if the allowed chemical shift range of its central carbon atom includes that observed chemical shift and if its signal multiplicity is the same as that of the observed signal.

The surviving ACFs are next tested for compatibility with the ^{1}H NMR data. HNMRPRUNE is similar to CNMRPRUNE in its function and operation. It also requires a database. For each ACF with a hydrogen-bearing central carbon atom, an allowed ^{1}H NMR chemical shift range is assigned to that (those) hydrogen(s). The allowed ^{1}H NMR chemical shift range of every such ACF surviving CNMRPRUNE must be matched by a signal somewhere in the observed ^{1}H NMR spectrum of the unknown or it is deleted. Signal multiplicity patterns are currently used to a limited extent by HNMRPRUNE in pruning ACFs.

If two-dimensional NMR data have been entered for the unknown, the substructural inferences made by INFER2D are passed to 2-DNMRPRUNE. Surviving ACFs within each carbon signal group whose central carbon connectivity conflicts with the signal connectivity inferred by INFER2D are deleted.

PRUNE also includes a routine to maintain the internal consistency among surviving ACFs. It does so by deleting wherever possible those ACFs causing the inconsistencies. The ACFs that survive this final step constitute the ACF shortlist. However, the chemist can then edit this ACF shortlist, thereby further reducing its size, by bringing to bear any structural information he or she may have from any source.

If PRUNE were perfect, the ACF shortlist would contain only those ACFs actually a part of the unknown structure. But the ACF is too small a fragment to permit a distinction to be made between each and every ACF solely on the basis of spectroscopic properties. Thus, in practice, the ACF shortlist will usually contain many more invalid ACFs—those not part of the unknown—than valid ones. Although the ACF shortlist should be as "short" as possible, the presence of invalid ACFs is not a major problem. Of course there is a major problem if PRUNE deletes a *valid* ACF; in that case the correct structure cannot be generated. For this reason PRUNE was designed on the principle that it is preferable to retain in invalid ACF on the ACF shortlist than to delete a valid one.

5.2.4. COCOA

The dual output of INTERPRET—the surviving ACFs that are the structural building units in structure generation and the collected substructural inferences that serve as the constraints on structure generation—is handed directly to COCOA (Fig. 5.3). It is important to recall that it is the exhaustiveness of the initial list of chemically stable ACFs that insures that no plausible structure will be excluded.

Very briefly, COCOA is a recently developed structure generator that represents a conceptual departure from currently available structure generating procedures.[2] It offers major advantages over these other systems.

1. COCOA uses all structural information produced by INTERPRET prospectively.
2. Its efficiency improves as the number of substructural inferences increases.

3. It can utilize potentially overlapping substructures without any preprocessing.

4. It can utilize alternative interpretations of the spectroscopic data prospectively and without any preprocessing.

5.3. PROBLEM SOLVING

5.3.1. Monochaetin

The data for the first problem are taken from the primary literature in natural products. Monochaetin is a fungal metabolite of molecular formula $C_{18}H_{20}O_5$. The structure assignment was reported in 1986[13] and depended heavily on two-dimensional NMR experiments, particularly long-range hydrogen–carbon correlations. This problem was of special interest in assessing the scope and limitations of INFER2D.

Table 5.2 is a tabulation of the one-dimensional ^{13}C and ^1H NMR data

TABLE 5.2. One-Dimensional ^1H and ^{13}C NMR Data
for Monochaetin

^{13}C NMR		^1H NMR			
Shift (ppm)	Mult.	Shift (ppm)	Integral	Mult.	Exch.[a]
205.94	S	6.79	1		
191.77	S	6.02	1		
169.10	S	5.29	1		
158.52	S	4.05	1		
145.52	S	3.76	1		
143.30	D	3.19	1		
116.22	S	2.13	3	S	
107.04	D	1.81	1		
105.73	D	1.48	1		
82.55	S	1.32	3	S	
52.13	D	1.11	3	D	
46.70	D	0.97	3	T	
43.66	D				
26.27	T				
19.49	Q				
18.92	Q				
14.39	Q				
11.45	Q				

[a] An "E" is entered if a signal disappears upon addition of D_2O.

reported in the paper and entered as input. The observed two-dimensional NMR data are shown in Table 5.3.

Input consists simply of the chemical shifts of coupled signals, element type, and the minimum and maximum number of intervening bonds. One-bond hydrogen–carbon correlations have been entered first; three-bond hydrogen–hydrogen correlations next; and last, all long-range hydrogen–carbon correlations that do not distinguish between two and three intervening bonds, i.e., each of the 27 long-range hydrogen–carbon correlations gives rise to two different interpretations. If unambiguous assignments were required, the set of 27 either-or statements would give rise to 2^{27}, or about 134 million, different sets of 27 unambiguous inferences. That number clearly conveys the degree of ambiguity in this information and the impracticability of solving this problem by generating each set of unambiguous inferences and treating each as a separate structure generation problem. But COCOA can contend with the large number of alternative inferences and use the structural information inherent in them prospectively, that is, efficiently.

Although the authors of the paper from which this problem was taken did deduce the presence of several substructures, these were not entered as user-defined substructural constraints in order to better test the full capability of INFER2D.

The only user interaction in this problem came in editing the ACF shortlist. Recall that PRUNE, which is responsible for creating the ACF shortlist, operates on the principle that it is preferable to retain an invalid ACF than to discard a valid one. Put differently: it is preferable for SESAMI to produce a broader range of alternative structures than to risk eliminating the correct structure. However, the user may choose to narrow the number of structures generated by excluding certain ACFs believed to be unlikely.

In running this problem, a user decision was made to assign the signals at δ 197 and 191 ppm solely to carbonyl carbon. Similarly, although the program assigned the signals at δ 147, 132, and 131 ppm to SP^2 or SP^3 carbon, the choice was limited to SP^2 carbon. Carbon atoms of SP hybridization were also excluded. In each case, these user decisions are compatible with assignments that would likely be made by chemists based on the available data, i.e., they are not high-risk decisions. But they are decisions that should be made by the chemist, not by the computer. Furthermore, such decisions are not intellectually demanding on the part of the chemist; they do not require extensive analysis of data.

SESAMI produced a single structure (Fig. 5.4) identical to that reported,[13] but used considerably less information than the authors of the paper. It is doubtful that a chemist could deduce this structure solely on the basis of the data used by SESAMI. Although this example casts the

TABLE 5.3. Two-Dimensional NMR Correlations for Monochaetin

Signal 1[a]	Signal 2[a]	Min[b]	Max[c]
C 143.30	H 6.79	1	1
C 107.04	H 6.02	1	1
C 105.73	H 5.29	1	1
C 43.66	H 3.76	1	1
C 52.13	H 4.05	1	1
C 46.70	H 3.19	1	1
C 26.27	H 1.81	1	1
C 26.27	H 1.48	1	1
C 11.45	H 0.97	1	1
C 19.49	H 2.13	1	1
C 18.92	H 1.32	1	1
C 14.39	H 1.11	1	1
H 4.05	H 3.76	3	3
H 3.19	H 1.11	3	3
H 3.19	H 1.81	3	3
H 3.19	H 1.48	3	3
H 0.97	H 1.48	3	3
H 0.97	H 1.81	3	3
C 18.92	H 3.76	2	3
C 19.49	H 6.02	2	3
C 26.27	H 1.11	2	3
C 43.66	H 1.32	2	3
C 43.66	H 4.05	2	3
C 46.70	H 1.11	2	3
C 52.13	H 3.76	2	3
C 82.55	H 1.32	2	3
C 82.55	H 3.76	2	3
C 82.55	H 5.29	2	3
C 105.73	H 6.02	2	3
C 107.04	H 2.13	2	3
C 107.04	H 5.29	2	3
C 116.22	H 3.76	2	3
C 116.22	H 5.29	2	3
C 116.22	H 6.02	2	3
C 116.22	H 6.79	2	3
C 143.30	H 3.76	2	3
C 145.52	H 6.02	2	3
C 145.52	H 6.79	2	3
C 158.52	H 2.13	2	3
C 158.52	H 6.02	2	3
C 158.52	H 6.79	2	3
C 169.10	H 4.05	2	3
C 191.77	H 1.32	2	3
C 205.94	H 1.11	2	3
C 205.94	H 3.76	2	3
C 205.94	H 4.05	2	3

[a] Element, chemical shift (ppm).
[b] Minimum number of intervening bonds.
[c] Maximum number of intervening bonds.

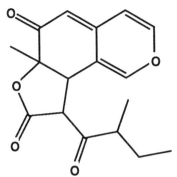

FIGURE 5.4. SESAMI output for the monochaetin
in problem.

early SESAMI program in a favorable light, more often than not in
problem solving, multiple structures result, sometimes too many to be
useful. In this particular problem, the two-dimensional NMR data are
especially information-rich and SESAMI is able to use them to good
advantage.

TABLE 5.4. One-Dimensional ^{1}H and ^{13}C NMR Data
for the Wasserman Compound

^{13}C NMR		^{1}H NMR			
Shift (ppm)	Mult.	Shift (ppm)	Integral	Mult.	Exch.[a]
197.73	S	7.70	1		
191.26	S	6.80	1	S	
166.98	D	6.60	1	S	
150.59	S	5.95	2	S	
147.03	S	5.55	1		
132.04	S	5.20	1		
131.42	D	5.05	1		
130.69	S	4.95	1		
119.50	T	4.05	1		
108.82	D	3.70	1		
108.67	D	3.40	1		
101.76	T	3.15	1		
96.27	D	2.85	1		
82.29	S	2.75	1		
49.01	T				
37.46	T				
32.08	T				

[a] An "E" is entered if a signal disappears upon addition of D_2O.

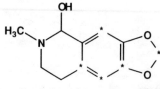

Prediction Accuracy = 95%

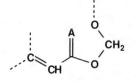

Substructure Input

FIGURE 5.5. INFERCNMR output for the Wasserman compound.

5.3.2. Wasserman Compound

The second problem is based on a compound of molecular formula $C_{17}H_{15}NO_4$ kindly provided by Professor Harry Wasserman of Yale University. One-dimensional ^{13}C and ^1H NMR data are shown in Table 5.4, the two-dimensional NMR data in Table 5.5. INFER2D again

TABLE 5.5. Two-Dimensional NMR Correlations
for the Wasserman Compound

Signal 1[a]	Signal 2[a]	Min[b]	Max[c]
C 166.98	H 7.70	1	1
C 108.82	H 6.80	1	1
C 108.67	H 6.60	1	1
C 101.76	H 5.95	1	1
C 131.42	H 5.55	1	1
C 119.50	H 5.20	1	1
C 119.50	H 5.05	1	1
C 96.27	H 4.95	1	1
C 49.01	H 4.05	1	1
C 49.01	H 3.70	1	1
C 37.46	H 3.40	1	1
C 32.08	H 3.15	1	1
C 37.46	H 2.85	3	3
C 32.08	H 2.75	3	3
H 7.70	H 4.95	3	3
H 5.55	H 5.20	3	3
H 5.55	H 5.05	3	3
H 5.55	H 3.40	3	3

TABLE 5.5. *(Continued)*

Signal 1[a]	Signal 2[a]	Min[b]	Max[c]
H 5.55	H 2.85	3	3
H 4.05	H 3.15	3	3
H 3.70	H 3.15	3	3
H 3.70	H 2.75	3	3
C 130.69	H 2.75	2	3
C 130.69	H 3.15	2	3
C 132.04	H 2.75	2	3
C 82.29	H 2.85	2	3
C 82.29	H 3.40	2	3
C 119.50	H 2.85	2	3
C 119.50	H 3.40	2	3
C 131.42	H 3.40	2	3
C 191.26	H 2.85	2	3
C 130.69	H 4.05	2	3
C 82.29	H 4.95	2	3
C 166.98	H 4.95	2	3
C 191.26	H 4.95	2	3
C 37.46	H 5.20	2	3
C 147.03	H 5.95	2	3
C 150.59	H 5.95	2	3
C 32.08	H 6.60	2	3
C 132.04	H 6.60	2	3
C 147.03	H 6.60	2	3
C 150.59	H 6.60	2	3
C 130.69	H 6.80	2	3
C 150.59	H 6.80	2	3
C 197.73	H 6.80	2	3
C 82.29	H 7.70	2	3
C 191.26	H 7.70	2	3
C 32.08	H 3.70	2	3
C 32.08	H 4.05	2	3
C 49.01	H 2.75	2	3
C 49.01	H 3.15	2	3
C 82.29	H 3.70	2	3
C 96.27	H 7.70	2	3
C 131.42	H 2.85	2	3
C 132.04	H 3.15	2	3
C 147.03	H 6.80	2	3
C 166.98	H 3.70	2	3
C 166.98	H 4.05	2	3
C 197.73	H 2.85		

[a] Element, chemical shift (ppm).
[b] Minimum number of intervening bonds.
[c] Maximum number of intervening bonds.

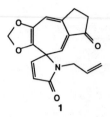
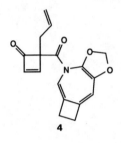
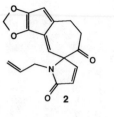
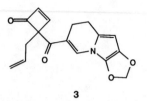

1

2

3

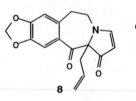
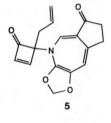

4

5

6

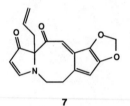
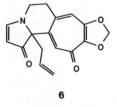

7

8

9

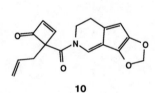

10

11

12

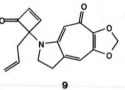
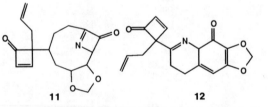
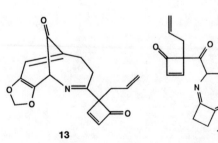
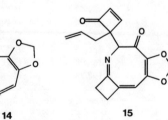

13

14

15

FIGURE 5.6. SESAMI output for the Wasserman compound.

played a central role in the structure elucidation. $^1H-^1H$ COSY, $^1H-^{13}C$ COSY, and long-range $^1H-^{13}C$ COSY data were used as in the monochaetin problem.

INFERCNMR was also applied to this problem. Using a matching tolerance of ±1.5 ppm, a single substructure was inferred at the 95% acuracy level. The output of INFERCNMR displays the substructure inferred embedded in the reference library entry from which it was retrieved (Fig. 5.5; predicted carbon atoms are shown as asterisks). That explicit substructure, including attached heteroatoms, was handed to COCOA. In addition, there was very limited and simple editing of the ACF shortlist as was done in the monochaetin problem.

SESAMI produced a total of 15 structures (Fig. 5.6). It now falls to the user, the specialist, to distinguish between the 15 structures. Some are readily eliminated, for example, the cyclobutenones, of which there are nine. It is important to note that in distinguishing among the 15 structures produced by SESAMI, the user has the assurance that no other structure equally compatible with the data has been overlooked. This is not always the case in assignments made in conventional structure elucidation.

Returning to the output of INFERCNMR once again (Fig. 5.5), we note that signals for two benzene ring carbons failed to match within the tolerance set. The chemist could decide to take added risk and assume the presence of those carbon atoms as well, and therefore a disubstituted dioxymethylene benzene fragment (which is, in fact, the case). If that fragment instead is handed to COCOA, a single structure (8), the correct structure, is reported by SESAMI. Thus, the interactive nature of SESAMI provides the user with considerable flexibility in problem solving.

5.4. CONCLUSIONS

The new strategy on which SESAMI is based offers considerable promise as a framework for computer-enhanced structure elucidation. The seamless link that the strategy permits between spectrum interpretation and structure generation, the ability of INFER2D to utilize the structural implications of all through-bond two-dimensional NMR correlations, even those that are ambiguous, the effective treatment of molecular symmetry, and the interactive nature of the program each significantly enhance SESAMI's power and versatility.

ACKNOWLEDGMENTS

The financial support of this research by the National Institutes of Health (GM37963), The Upjohn Company, and Sterling Drug, Inc., is gratefully acknowledged. Co-workers M. Badertscher, B. D. Christie, M. Clay, M. Farkas, A. Lipkus, C. A. Shelley, and V. K. Velu deserve much of the credit for the work described herein.

REFERENCES

1. M. E. Munk, C. A. Shelley, H. B. Woodruff, and M. O. Trulson, *Fresenius Z. Anal. Chem.* **313**, 473 (1982).
2. B. D. Christie and M. E. Munk, *J. Chem. Inf. Comput. Sci.* **28**, 87 (1988).
3. H. Kessler, M. Gehrke, and C. Griesinger, *Angew. Chem., Int. Ed. Engl.* **27**, 490 (1988).
4. B. D. Christie and M. E. Munk, *Anal. Chem. Acta* **200**, 347 (1987).
5. C. A. Shelley, T. Hays, M. E. Munk, and R. V. Roman, *Anal. Chim. Acta* **103**, 121 (1978).
6. R. E. Carhart, D. H. Smith, N. A. B. Gray, and C. Djerassi, *J. Org. Chem.* **46**, 1708 (1981).
7. H. Abe, I. F. Okuyama, and S.-I. Sasaki, *J. Chem. Inf. Comput. Sci.* **24**, 220 (1984).
8. W. Bremser and W. Fachinger, *Magn. Reson. Chem.* **27**, 1056 (1985).
9. M. Carabedian, I. Dagane, and J. E. Dubois, *Anal. Chem.* **60**, 2186 (1988).
10. A. H. Lipkus and M. E. Munk, *J. Chem. Inf. Comput. Sci.* **28**, 9 (1988).
11. C. A. Shelley and M. E. Munk, *Anal. Chem.* **54**, 516 (1982).
12. J. Neter, W. Wasserman, and M. H. Kutner, *Applied Statistical Models*, Chap. 10, Irwin, Homewood, Illinois (1985).
13. P. S. Steyn and R. Vleggaar, *J. Chem. Soc., Perkin Trans. 1*, 1975 (1986).

Computer-Assisted Mass Spectral Interpretation: MS/MS Analysis

6

Kevin J. Hart and Chris G. Enke

6.1. INTRODUCTION

Computer-assisted interpretation of molecular spectra has been, and continues to be, one of the foremost challenges to chemists interested in the elucidation of molecular structure. The driving forces spurring research in this area include not only the incumbent advantages of a computerized interpretation of a spectrum (e.g., speed and consistency), but also the nature and volume of the data produced by modern, computer-controlled analytical instrumentation. For some recent instruments, development of a computerized "expert" interpretive system using a traditional knowledge-engineering approach is impractical for the simple reason that the required human expert capable of synthesizing large quantities of multidimensional spectral information does not exist. We have developed a new approach to computer-assisted structure elucidation based on computer pattern recognition and automated rule generation. This approach has been applied in the interpretation of mass spectrometry/mass spectrometry (MS/MS) spectra.

Kevin J. Hart • Oak Ridge National Laboratory, Oak Ridge, Tennessee, 37831-6365. Chris G. Enke • Department of Chemistry, Michigan State University, East Lansing, Michigan 48824.

Computer-Enhanced Analytical Spectroscopy, Volume 3, edited by Peter C. Jurs. Plenum Press, New York, 1992.

6.1.1. Structure Elucidation Using MS/MS Spectra

The mass spectrometry/mass spectrometry technique has played an important role in solving a variety of chemical analysis problems in academic, industrial, and governmental laboratories. This technique has provided new and useful data in studies of the environment, natural products, industrial products, foods, forensic science, petroleum products, bioorganic compounds, and pharmaceuticals.[1, 2] The MS/MS technique was initially popular for its potential in analyzing mixtures but has increasingly been applied to structure elucidation.[3] Interest in automating the structure elucidation process has grown with the recognition of its applicability.

A typical MS/MS experiment begins by acquiring a primary mass spectrum. Ions formed in the mass spectrometer ion source and appearing at specific mass-to-charge ratios (i.e., m/z) in the primary mass spectrum are then individually selected by an initial mass filter and introduced into a "collision cell" pressurized with a neutral target gas (often argon). The

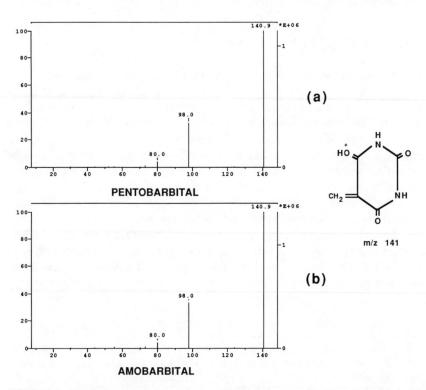

FIGURE 6.1. Daughter spectra obtained on a TQMS instrument for (a) pentobarbital and (b) amobarbital using single collision conditions (reprinted with permission from Ref. 12).

selected ions then undergo collision-induced dissociation (CID) to form fragment ions. The collision products (fragment ions) are then introduced into a second mass filter so the m/z ratios for the products can be recorded. The mass spectrum of the CID products is referred to as a daughter spectrum. The daughter spectra of the m/z 141 ions of pentobarbital and amobarbital are shown in Fig. 6.1. Since these spectra are identical, it is likely that the m/z 141 ion results from a substructure common to the two compounds. The aim of this work is to exploit similarities in the daughter spectra of known compounds for inferring the presence (or absence) of substructures in unknown compounds. The identification of multiple substructures of a molecule can lead to its complete characterization.

There are a great many different tandem mass spectrometers capable of providing MS/MS spectra that vary in the way mass analysis and CID are performed.[1] Other MS/MS instruments also exist where the ionization, dissociation, and product analysis steps are separated in time rather than space.[4] The data shown in this discussion are all derived from a triple quadrupole mass spectrometer (TQMS).[5] The full MS/MS data space (the MS/MS map) consists of a primary mass spectrum and a daughter spectrum for each of the m/z values in the primary mass spectrum. The information contained in all of the MS/MS scans (parent, daughter, neutral loss) on instruments, such as the TQMS, resides in this data space. The problem facing chemists is to extract the structurally relevant information from these data and use it to deduce the molecular structure of unknowns. This problem can be quite challenging, since the size of the MS/MS data space increases much more rapidly with mass than the MS data space.[6] While much has been done to provide sophisticated data acquisition systems,[7-9] data interpretation software that fully exploits the information obtainable from MS/MS instruments has, heretofore, not been realized.[10]

We describe here the approach adopted in our laboratory to solve the problem of automating the identification of molecular structure from MS/MS spectra. This approach is based on the development of rules which can be used to detect the presence (or absence) of various substructures followed by exhaustive generation of candidate molecular structures. The size of the spectral database required by this approach to provide a "training set" for the rule generation software is relatively small compared to the database needed for spectral matching. A significant advantage of compound identification by substructure characterization versus full spectrum matching is that the exact compound to be identified need not have been included in the training set. The ultimate success of this approach depends on the number and accuracy of the rules used to make predictions of the presence of substructures in unknowns. Consequently, most of this discussion is devoted to the newly developed rule generation software and an evaluation of the rules produced by this software. While the TQMS instrument will be used to illustrate the concepts in this paper, many of these ideas are equally applicable to other instruments capable of producing MS/MS data.

6.1.2. The Automated Chemical Structure Elucidation System

The Automated Chemical structure Elucidation System (ACES) under development in our laboratory generates candidate molecular structures for unknown compounds.[6, 10-12] The system is based on the ability to use MS/MS data to determine the molecular formula and molecular substructures present in or absent from the molecular structure of an unknown compound. This information is then used to constrain the exhaustive generation of candidate structures. The ACES approach follows the DENDRAL concept, where interpretive programs for MS and NMR data were used to identify substructure constraints for use in a structure generator.[13] ACES, however, uses a distinctly different method for obtaining substructure identification rules based on MS/MS data and is automated rather than interactive. The system has been designed to allow the eventual incorporation of the ability to recommend ancillary experiments which are most likely to resolve ambiguous results.

The major program components and the two distinct operational modes of the system are shown in Fig. 6.2. The first operational mode, following the dashed lines in the figure, is the learning mode. This is the mode of operation used to create the substructure identification rules. The first step involves the creation of a reference database containing the full MS/MS map for reference compounds of known structure. The reference database is used by the MAPS (Method for Analyzing Patterns in Spectra) software, developed in this laboratory, to generate substructure identification rules.[14-16] There are two types of rules: inclusion rules and exclusion rules. An inclusion rule can predict the presence of a substructure while an exclusion rule can predict the absence of a substructure. These rules are stored and used in the identification mode of the ACES software. The identification mode, shown in Fig. 6.2 with a solid line, begins with the acquisition of the MS/MS map for the unknown compound. The substructure identification rules are then applied to the MS/MS data obtained. The result is a list of substructures present in or absent from the structure of the unknown compound. Molecular formula information is provided by the MFG (Molecular Formula Generator), a program developed in this laboratory. The molecular formulas and the substructure list are used by the structure generator to provide candidate structures for the unknown compound. The next three sections introduce the ACES modules that

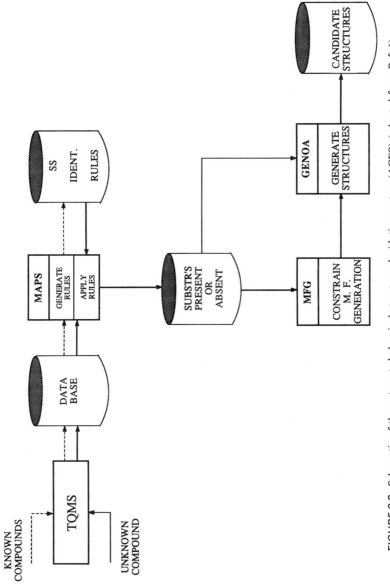

FIGURE 6.2. Schematic of the automated chemical structure elucidation system (ACES) (adapted from Ref. 6).

generate candidate molecular formulas, substructure identification rules, and molecular structures.

6.1.2.1. Generation of Molecular Formulas

The molecular formula required by the structure generator can be deduced from information in the MS/MS data space using a tandem mass spectrometer with unit mass resolution or by measurement of the exact mass of the molecular ion using a high-resolution mass spectrometer. This piece of information may not be present in electron impact mass spectra so a chemical ionization (CI) experiment, a "soft ionization" technique that often yields molecular ions, may be necessary. The molecular weight and the elemental compositions of the substructures found to be present in the structure of the unknown compound are input into the MFG program.[17] The MFG program is an exhaustive molecular formula generator that uses the nominal mass of a molecule and elemental constraints to provide a list of candidate molecular formulas.

It is very desirable to reduce the number of candidate molecular formulas to one, since a separate structure generation session must be performed for each possible molecular formula. Information exists in MS/MS spectra which can help achieve this goal. A program has been developed which uses the ratios of daughter ion intensities found in the daughter spectrum of the $M + 1$ ion to determine the number of carbon atoms in the molecular ion.[17] Several advances in this technique have been recently reported. Bozorgzadeh[18] has recently described further development and generalization of this method for the determination of all the constituent elements in daughter ions. Glish *et al.*[19] have also demonstrated the utility of additional stages of mass spectrometry (e.g., MS/MS/MS) for further resolving the elemental composition of daughter ions. Currently, the MFG program in our laboratory only utilizes the number of carbon atoms determined by ^{13}C ratio measurements in the $M + 1$ daughter spectrum. The exact mass of the molecular ion of the unknown, if available, can also be input to the MFG program to unambiguously generate the molecular formula of the unknown.

6.1.2.2. Generation of Substructure Identification Rules

The MAPS software automatically generates the substructure identification rules for ACES and has been recently modified in the latest version to increase the reliability and recall of rules through the use of "feature-combinations."[16] The reliability and recall of a rule are given by Eqs. (6.1) and (6.2):

$$REL = \frac{\# \text{ of correct predictions}}{\text{total } \# \text{ of predictions}} \qquad (6.1)$$

$$REC = \frac{\# \text{ of correct predictions}}{\text{total } \# \text{ of possible correct predictions}} \qquad (6.2)$$

where REL is the rule reliability and REC is the rule recall.

The goal of the ACES system is to obtain a single, definite structure for an unknown or, at the very least, a set of candidate structures which is consistent with the structural information contained in the MS/MS spectra of the unknown. While 100% rule reliability is required to ensure that this goal is achieved, 100% recall is not an absolute requirement. A subset of all the possible substructure identifications can lead to a single structural candidate for an unknown. However, the identification of a larger number of the substructures contained in an unknown compound, as will generally be the case for a rulebase consisting of rules with high recall, will increase the probability that only a single structure will be obtained for an unknown.

In previous versions of MAPS,[14, 15] a substructure identification rule had the form:

"IF $\langle$spectral feature $f_1\rangle$ is present
and $\langle$spectral feature $f_2\rangle$ is present
and $\langle$spectral feature $f_x\rangle$ is present
THEN substructure 'X' is present"

A "match factor" (MF) was then used to specify the minimum fraction of rule clauses that had to be true (i.e., the minimum number of spectral features which had to be found in the MS/MS spectra of an unknown) for a substructure prediction to be made by the system. The disadvantage of this method of rule generation and rule application is that any combination of features (greater than or equal to the number specified by MF) present in the MS/MS spectra of an unknown will cause a substructure prediction to be made by the system. Using this method, greater rule reliability could only be had at the expense of recall. Thus, a new method was developed to reduce the likelihood of a false substructure prediction and to increase rule recall. This method is based on specific combinations of spectral features (i.e., "feature-combinations) and has been shown to provide optimal reliability *and* recall.[16]

A "feature-combination" is a set of features which collectively have a uniqueness (with respect to the reference database) greater than or equal to any individual feature in the set for a particular substructure. A rule based on "feature-combinations" has the general form:

"IF $\langle$ spectral features $f_{a1}, f_{a2}, ...\rangle$ are present
or $\langle$ spectral features $f_{b1}, f_{b2}, ...\rangle$ are present
or $\langle$ spectral features $f_{n1}, f_{n2}, ...\rangle$ are present
THEN substructure 'X' is present."

A substructure identification is made using this type of rule if any of the rule clauses is true (i.e., the features in a feature-combination are found in the MS/MS spectra of an unknown) since each feature-combination has 100% uniqueness, and therefore 100% reliability, for the specified substructure. The feature-combination rule obtained for the phenothiazine substructure is shown in Fig. 6.3. This rule has 100% reliability and 100% recall for the phenothiazine substructure *with respect to the reference database*. Five correct predictions were obtained when this rule was used to analyze the MS/MS spectra of 20 test compounds (compounds not in the reference database), 5 of which contained a phenothiazine substructure.[20] Thus, in this test, the MAPS feature-combination rule for the phenothiazine

IF	" PD (198.0 154.0) **and**	D (198.0)	[100,92] "
OR	" PD (197.0 196.0) **and**	PD (198.0 154.0)	[100,76] "
OR	" PD (197.0 196.0) **and**	PD (197.0 153.0)	[100,76] "
OR	" PD (197.0 196.0) **and**	PD (196.0 152.0)	[100,76] "
OR	" PD (198.0 171.0) **and**	PD (198.0 154.0)	[100,76] "
OR	" PD (197.0 153.0) **and**	PD (196.0 152.0)	[100,76] "

THEN substructure *phenothiazine* is present.

REL = 100% and REC = 100% with respect to reference database

U_i = 40%, C_i = 70%, Cc = 70%, mass filter enabled

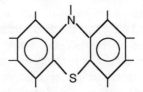

phenothiazine substructure

FIGURE 6.3. MAPS feature-combination rule for the phenothiazine substructure (reprinted with permission from Ref. 20).

substructure also had 100% reliability and 100% recall *with respect to nonreference compounds.*

6.1.2.3. Generation of Molecular Structures

The molecular formulas and the substructures found to be present in or absent from the structure of the unknown compound are used as constraints in GENOA, an exhaustive structure generation program.[21] This program was developed as part of the DENDRAL project. GENOA has been modified for use in ACES to provide automated generation of candidate structures. The GENOA software also includes a structure search program, STRCHK. This program provides a list of substructures contained in the structures of the reference compounds, which is used by the MAPS software in generating the substructure identification rules.

One of the key issues in using this approach is the accuracy of the input constraints to GENOA. Each of the candidate structures that GENOA produces is completely consistent with the input constraints. If any of the input constraints is wrong, then all of the candidate structures are wrong. This fact leads to the paramount importance of accurate substructure and molecular formula determinations and accounts for the continuing search for better ways to generate reliable substructure identification rules.

6.1.3. Recommendation of Ancillary Experiments

In general, systems which use structure generators provide a method for ranking candidate structures to assist the user in determining the most likely candidate structure for the structure of an unknown compound. The approach to be used in a forthcoming version of ACES seeks to assist the user in identifying ancillary experiments which will potentially reduce the number of candidate structures.[6] Implementation of this approach will establish a "feedback loop" which includes the instrument, the TQMS in this case. This approach preserves the experimental versatility that MS/MS instrumentation can provide to the structure elucidation chemist. The components necessary to implement this approach are shown in the schematic provided in Fig. 6.4. When multiple candidate structures are produced by the first experiment, a substructure analysis of the candidate structures is performed by a modified version of STRCHK.[22] The original version of STRCHK was also developed as part of the DENDRAL project. The modifications made to STRCHK provide the level of automation necessary for integration of this program into ACES. This program outputs a list of substructures that differentiate among the candidate structures (discriminating substructures). MAPS rules that have been collected under different

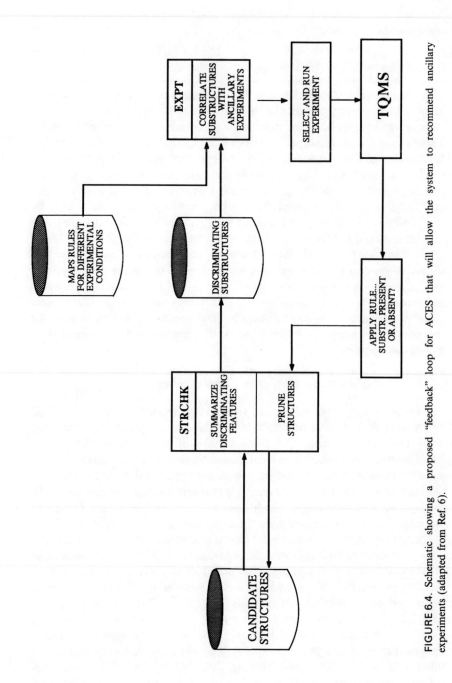

FIGURE 6.4. Schematic showing a proposed "feedback" loop for ACES that will allow the system to recommend ancillary experiments (adapted from Ref. 6).

operating conditions (e.g., a higher collision gas pressure to provide more collisions) could then be consulted by the EXPT program (currently under development) to determine if a new experiment exists that can confirm the presence of the discriminating substructures. To implement this feature of ACES, multiple rulebases must be generated using data obtained under a variety of conditions. Once the EXPT program has selected the conditions for a new experiment, the new instrumental conditions can be passed to the user (or directly to the instrument control software in a totally automated system) to be implemented. The list of candidate structures can then be pruned depending on the results of the new experiment. This cycle can then be repeated until the list of candidate structures has been reduced to one, the structure of the unknown compound, or the list of discriminating experiments has been exhausted.

MS/MS databases for reference compounds acquired under different operating conditions are being compiled to determine the optimal operating conditions for use in generation of substructure identification rules. The variables that may be explored include ionization conditions (e.g., electron impact, chemical ionization, FAB), ion charge (e.g., positive ions or negative ions), and collision conditions (e.g., collision energy, collision gas pressure, target gas, reactive collisions). It is not expected that one set of instrumental conditions will be optimal for all substructures but rather a number of instrumental conditions will be found which are best suited for some fraction of the substructures being investigated.

6.2. THE MS/MS SPECTRA INTERPRETATION SOFTWARE: MAPS

The new version of MAPS has been divided into several C program modules and is currently running on a DEC VAXstation 3200 computer.[16] The main program modules are GENT (GENerate Training set), the MAPS program which generates feature-combination rules, and the RULE program which applies rules to MS/MS data of unknown compounds. The major inputs, outputs, and interactions of these programs are shown in the schematic provided in Fig. 6.5. The GENT program pre-processes several types of data to create a spectral feature and substructure array (the training set). The MAPS program then generates substructure identification rules from the training set using a number of program parameters to control the generation of rules. The program then writes the rules to a file for subsequent use by the RULE program. The RULE program applies rules to the MS/MS data of an unknown compound and displays the names of the substructures identified as present in the unknown. These substructures are also written to a results file. The RULE results file, in turn, is used

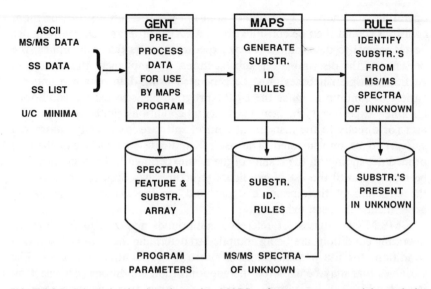

FIGURE 6.5. Schematic showing the major MAPS software program modules and the requisite input to and output from the programs.

by ACES to provide structural constraints to the automated structure generator.

6.2.1. Creation of the Training Set

The GENT program begins by prompting for the primary mass spectrum and daughter spectra filenames of all the reference compounds. A reference name for each compound is also requested. MS/MS spectral features are then identified within all the MS/MS spectra and tabulated for future reference. Each spectral feature is followed by a list of compound names which contain the feature, so the final result is very much like an inverted database. The MS/MS spectral features include primary ions, "P (m/z)," daughter ions "D (m/z)," neutral loss "NL (amu)," and parent–daughter pairs "PD $(m/z\ m/z)$." Three user-definable intensity thresholds, PTHRESH, DTHRESH, and PDTHRESH, can be used to limit the spectral features included in the training set. The PTHRESH and DTHRESH parameters limit the primary and daughter scan ions to those that exceed the threshold value. The PDTHRESH parameter changes the threshold for parent ion selection. For example, a 1% threshold was used during data acquisition to select ions from the primary mass spectrum for collision-induced dissociation and subsequent acquisition of daughter spectra. The GENT user can select a higher threshold to remove daughter spectra of low-intensity parent ions. Typical values for PTHRESH, DTHRESH, and PDTHRESH are 0.1%, 1%, and 1%, respectively. Another program parameter, MINWF, defines a minimum number of compounds, typically 3 compounds, in which a spectral feature must be found before the feature is included in the training set. The current reference database contains 100 compounds, many of which are regulated drugs.

After the data files for each reference compound have been entered, the GENT program prompts for the substructure information. This information is saved in a file as a list of substructures contained in each reference compound. The ASLS (Automated Structure Library Search) program is a modified version of the STRCHK program originally developed at Stanford University as part of the DENDRAL project.[13] The source code for the commercial version of this software was modified to check all the structures for the reference compounds against a substructure library. Thus, the connectivity table for all reference compounds must be available. GENOA was used to define all the reference compounds used for this work. The current substructure library, defined for use with the reference compounds, contains 161 substructure definitions. Of these substructures, 121 are represented in at least one compound in the reference database. The name of a file containing additional element and substructure information is then requested. This file contains 19 element definitions that include important isotopes and relevant oxidation states. These definitions consist of the mass and valence of these elements. The mass and elemental composition of the 161 substructures in the substructure library are also included. The current software only utilizes the substructure masses (i.e., in the MAPS program) but the other substructure information may prove useful in subsequent versions of this software.

Another set of parameters also limits the spectral features included in the training set. These parameters are a minimum feature uniqueness (U) and a minimum feature correlation (C). Equations (6.3) and (6.4) are used to calculate U and C values for spectral features and for feature-combinations:

$$U_X = \frac{\text{number of compounds with } SS_X \text{ and } F_X}{\text{number of compounds with } F_X} \tag{6.3}$$

$$C_X = \frac{\text{number of compounds with } SS_X \text{ and } F_X}{\text{number of compounds with } SS_X} \tag{6.4}$$

where U_X is the spectral feature uniqueness, C_X is the spectral feature correlation, SS_X is substructure X, and F_X is spectral feature X.

Typical starting values for U and C are 10%. Thus, all spectral features in the training set have at least 10% uniqueness and correlation with respect to at least one substructure in the substructure library. The actual U and C values for each feature in the training set that exceed the specified U/C minima are included in the training set file.

The GENT program concludes by prompting for the name of the training set file. An excerpt from a training set file generated using 40 compounds and 2 substructures is provided in Fig. 6.6. The file begins with a list of the defined substructures and the mass of each substructure. The next section of the file contains lists of the name of each reference compound and the substructures found in the compound. The last section is a list of spectral features that meet the threshold, MINWF (Minimum number of compounds With Feature) and the U/C criteria. Each spectral feature is followed by a series of bits which identify the reference compounds that yield the feature. Since 40 compounds were used for this example, there are 40 bits in the string following each spectral feature in Fig. 6.6. The list of U and C values follows the bit string for each feature. GENT required approximately 30 minutes of CPU time to process 100 primary and 5749 daughter spectra of 100 reference compounds. The GENT output file contains all of the spectral and substructure data required by the MAPS program. The MAPS program described in the next section generates feature-combinations without regard to the origin of the spectral data. Thus, it should be possible to generate rules based on IR, NMR or even combined IR, NMR, and MS/MS data using the existing MAPS

```
SS18 80.0
SS50 96.0

GMR10        SS18 SS50
GMR11        SS18 SS50
GMR12        SS18 SS50
...
(and so on for a total of 40 compounds)

(PD 207.0 192.0)       0001000010000000011000100000000100000000
(PD 190.0 189.0)       1010001000000000000000000000101000010000
(PD 185.0 184.0)       0011000001000000000000000000000100100000
(PD 207.0 179.0)       0001001010000000001000000000000100000000
...
(NL 77.0)    0100000101000010111000000000100000011111
(D 77.0)     1111101111111011111000111000111110111111
(P 77.0)     1111111111111011111111111111111111111111
...
(and so on)
```

FIGURE 6.6. An excerpt from a GENT file showing the data format used by the MAPS program (reprinted with permission from Ref. 16).

program. Feature recognition software for IR and NMR data will have to be incorporated into the GENT program, however, before this experiment may be attempted.

6.2.2. Search Strategies for Generation of Feature-Combinations

The latest version of the MAPS program includes several enhancements not found in previous versions of MAPS. The most important of these enhancements is the ability to rapidly generate rules based on feature-combinations. Two important strategies were employed in the new software to avoid the computation "explosion" which grows rapidly with an increasing number of starting elements. For example, exhaustive generation of all combinations of the 964 spectral features that have at least 10% uniqueness and correlation for the barbiturate substructure would result in $2^{964} - 1$ different feature-combinations. Uniqueness and correlation values for each of these feature-combinations would have to be calculated with respect to the training set to determine if they should be included in a substructure identification rule. The computation time required to complete this operation would be measured in years. Obviously, strategies had to be developed to limit the size of the feature-combination search space.

6.2.2.1. Pruning the Feature List

The first strategy used in the new software is to limit the initial number of spectral features passed to the feature-combination generator by specifying a minimum initial uniqueness (U_i) and correlation (C_i) value higher than those specified in the GENT program. Higher values of U_i and C_i are usually necessary for reliable rule generation. The U/C values used by GENT were kept low to allow several different sets of U_i and C_i values to be explored within MAPS without having to regenerate the training set. The number of initial features obtained for any substructure decreases with increasing U_i and C_i values.[16] For example, 56 features were obtained for the barbiturate substructure using a value of 30% for U_i and C_i (down from the 964 features with at least 10% U_i and C_i). The number of features selected for the barbiturate substructure can be further reduced with higher values of U_i or C_i. The number of features required to optimize rule recall, however, also must be considered.

Different sets of U_i and C_i values can also change the nature of the spectral features included in the initial feature list. For example, there are two sets of U_i/C_i values that yield 11 spectral features for the barbiturate substructure (i.e., 20%/80% and 70%/30%). In the first set of values, U_i is low and C_i is high, while the opposite is true for the second set of values. Almost all (10 out of 11) of the spectral features selected using the low U_i

and high C_i values are primary, daughter, and neutral loss features. These features tend to be more general than parent ion–daughter ion pairs and thus have lower U_i values. However, the features selected when U_i was set high and C_i low were only parent ion–daughter ion pairs, which tend to be more specific than the other types of spectral features observed in MS/MS spectra and thus tend to have high U values.

6.2.2.2. Pruning Feature-Combination Search Branches

A second strategy used in the new software provides early detection and removal of nonproductive search branches during rule generation. A feature-combination is constructed by adding features from the list of initial features until no false positives are observed (i.e., $U = 100\%$) or the recall for the combination falls below a specified minimum value, C_c. A search branch is terminated when the correlation of a candidate feature-combination falls below the C_c value or when the compound list associated with a candidate feature-combination is comprised of compounds already identified a sufficient number of times by previously generated feature-combinations (i.e., when the HITS limit is met). Additional computational efficiency is obtained using the new MAPS algorithm which eliminates false positives (an integer value) rather than increasing uniqueness (a floating-point value). Integer operations can be performed on the computer system used for this work more rapidly than floating point operations.

The new MAPS algorithm is illustrated in Fig. 6.7, which shows the generation of candidate feature-combinations for the *barbiturate* substructure from a list of 23 initial spectral features. These spectral features were selected using $U_i = 35\%$ and $C_i = 30\%$. The minimum feature-combination correlation, C_c, was set to 30%. The first feature-combination shown in Fig. 6.7 was simply the first feature in the feature list. The second feature-combination was comprised of two features from the single feature list with a combined uniqueness of 100% and correlation of 30%. This feature-combination was included in the barbiturate rule, since it had 100% uniqueness for the barbiturate substructure and met the C_c value. An overall recall value was also printed by the software for feature-combinations included in a rule as shown for the second feature-combination in Fig. 6.6.

The fate of each feature-combination can be determined by examining the "tag" that the MAPS software associates with each candidate feature-combination. The "more" tag, for example, was used to indicate that a feature-combination had associated false positives and required additional spectral features (e.g., feature-combinations #1 and #11 in Fig. 6.7). The remainder of these tags were used to identify the reason that a search branch was terminated. The "low" tag was used to indicate termination of a search branch because the correlation of a combination fell below the limit set by

```
1  (PD 98.0 80.0 P1)  U=85% C=46% more
2  (PD 98.0 80.0 P1) (PD 169.0 97.0 P1)  U=100% C=30% 30% recall
3  (PD 98.0 80.0 P1) (PD 98.0 27.0 P1)  U=100% C=38% 46% recall
4  (PD 98.0 80.0 P1) (PD 98.0 28.0 P1)  U=100% C=38% 46% recall

...

5  (PD 98.0 80.0 P1) (D 16.0 P1)  U=100% C=7% low
6  (PD 98.0 80.0 P1) (PD 98.0 70.0 P1)  U=85% C=46% worse
7  (PD 98.0 80.0 P1) (PD 160.0 104.0 P1)  U=0% C=0% low

...

8  (PD 98.0 27.0 P1)  U=83% C=38% hits
9  (PD 98.0 28.0 P1)  U=75% C=46% more
10 (PD 98.0 28.0 P1) (PD 169.0 126.0 P1)  U=100% C=23% low
11 (PD 98.0 28.0 P1) (PD 98.0 44.0 P1)  U=100% C=38% hits

...

12 (PD 54.0 27.0 P1) (NL 111.0 P1)  U=71% C=38% more
13 (PD 54.0 27.0 P1) (NL 111.0 P1) (PD 97.0 55.0 P1)  U=80% C=30% hits
14 (PD 54.0 27.0 P1) (NL 111.0 P1) (PD 69.0 39.0 P1)  U=100% C=38% back
```

FIGURE 6.7. Selected candidate feature-combinations generated by the MAPS program during rule generation (reprinted with permission from Ref. 16).

the C_c parameter. The addition of features to any subsequent feature-combinations would not be productive, since the correlation would either remain the same or decrease from the current value. The "worse" tag was used to indicate that addition of the last spectral feature did not decrease the false positives associated with the feature-combination (e.g., feature-combination # 6). The "hits" tag indicates that the feature-combination does not identify any reference compounds that have not already been identified a sufficient number of times in previously generated feature-combinations as specified by the HITS parameter (e.g., feature-combination # 8). The last tag used by MAPS, "back," indicates that a redundant feature was detected in the feature-combination (e.g., feature combination # 14). For feature-combination number 14 in Fig. 6.7, removal of the second feature of the combination results in a feature-combination which still has 100 % uniqueness, since the combination of the first two features has only 71 % uniqueness (see feature-combination # 12). This operation is performed for each candidate feature-combination with 100 % uniqueness when the BACK-CHECK parameter is enabled. The objective of using this parameter is to eliminate multiple pathways to the same feature-combination. Optimal system performance varies, but generally BACKCHECK should be enabled only when it is expected that large numbers of feature-combinations will be generated.

6.2.3. Using MAPS with Feature-Combinations

Sample output for the MAPS program is provided in Fig. 6.8. The program begins with a prompt for the training set filename. The program then provides a summary of the training set that includes the number of compounds, substructures, and spectral features in the training set. The primary and daughter scan conditions are also provided. These conditions can vary for different training sets (e.g., EI or NCI for primary scan features and single or multiple collision for daughter scan features). The substructure reference names are also provided as they are used to identify the substructures in the program.

                MAPS
Method for Analyzing Patterns in Spectra
Training set file: **GENT_P1.OUT**
100 compounds   121 substructures
Primary conditions: ""
Daughter conditions: P1
4386 features

Substructures: SS158 SS159 SS157 SS37 SS156 SS36 SS35 SS131 (...)

MAPS> **SINGLEU .4**
MAPS> **SINGLEC .7**
MAPS> **SINGLE SS132**

SS132 10 features   3 s

MAPS> **USORT**
MAPS> **NLFILTER**
MAPS> **FILTER**
MAPS> **PFILTER**
MAPS> **PRINT**

1: (PD 198.0 154.0 P1)   U=92 C=92
2: (PD 197.0 153.0 P1)   U=90 C=76
3: (PD 196.0 152.0 P1)   U=90 C=76
4: (PD 198.0 171.0 P1)   U=76 C=76
5: (PD 197.0 196.0 P1)   U=64 C=84
6: (PD 70.0 27.0 P1)   U=45 C=84
7: (D 198.0 P1)   U=41 C=92

MAPS> **DELETE 6**

MAPS> **COMBOC .7**
MAPS> **COMBINATION SS132.RUL**

SS132   7 features   13 compounds   6 clauses 100% recall

FIGURE 6.8. Output produced by the MAPS program during rule generation for the phenothiazine substructure.

The MAPS software is command-line driven. The first step in using the code is the creation of a feature list. This operation is accomplished using the SINGLEU, SINGLEC, and SINGLE commands as shown in Fig. 6.8 (i.e., a single feature uniqueness and correlation of 40% and 70%, respectively, for the SS132 substructure). The MAPS software then reports the number of spectral features which meet these values. The feature list can be further manipulated by sorting the list by uniqueness (USORT), correlation (CSORT), or mass (MSORT). There are several user selectable mass filters implemented in the new MAPS program. The mass filter, called FILTER, removes all primary, daughter, and neutral loss features with masses larger than the nominal mass of the substructure. The NLFILTER mass filter removes nonintegral neutral losses which arise in the reference database from the presence of doubly charged ions in the reference spectra and occasional inaccurate mass assignments. Another mass filter (PDFILTER) removes all parent ion–daughter ion pairs that represent a neutral loss of a mass larger than the nominal mass of the substructure specified by the SINGLE command. A mass filter, PFILTER, is also available that removes parent ion–daughter ion pairs with a parent mass greater than the nominal mass of a given substructure. The feature list can be displayed using the print command as shown in Fig. 6.8. The features in this list are numbered to allow the deletion of undesirable features using the MAPS DELETE command. All daughter, neutral loss, and parent ion–daughter ion pairs are labeled by a tag indicating the collision conditions used to acquire the reference spectra (e.g., P1: single collision conditions).

The minimum acceptable feature-combination correlation, C_c, is specified using the COMBOC command as shown in Fig. 6.8. Feature-combinations can then be generated by using the COMBINATION command and entering a rule filename. Default values for all pertinent program parameters are stored in an initialization file that is read when the MAPS program is invoked. A summary of the feature-combinations generated is displayed to the terminal after each COMBINATION command is executed. In this example, 6 feature-combinations (or rule clauses) were obtained for SS132 with an overall recall of 100% with respect to the reference database. When the program parameter VERBOSE is enabled, each feature-combination that is checked is displayed to the screen. This parameter was enabled to provide the text for Fig. 6.7. Only those feature-combinations that have 100% uniqueness for the indicated substructure, however, are written to the rule file.

6.2.4. Application of the MAPS Rules to Unknowns

The RULE program compares the feature-combinations found in a rule file to those present in the MS/MS spectra of an unknown. Three filenames are required to analyze an unknown with the current version of MAPS (i.e., the rule, primary scan, and daughter scan filenames). The RULE program lists the number of rule clauses (feature-combinations) that "are true for the unknown and the total number of clauses in the rule. A sample session is provided in Fig. 6.9, where MAPS rules for the phenothiazine and barbiturate substructures were used to analyze 20 test compounds (i.e., compounds not present in the reference database). The total time required to load the MS/MS spectra of the test compounds and to apply the two MAPS rules was approximately 3 minutes. When the VERBOSE parameter is enabled in the MAPS initialization file, the rule clauses that are true for an unknown are displayed to the terminal so the user can examine the spectral features in the rule clauses. A RULE

RULE - checks hits, reliability, and recall of MAPS rules
Rule file[*.rul]: **SS132_TEST**
Rule file[*.rul]: **SS143_TEST**
Rule file[*.rul]: **<CR>**
Primary conditions: ""
Daughter conditions: P1
Compound name: **UNK1**
Primary scan "" [*.dat]: **khunk1mp**
Daughter scan P1 [*.dat]: **khunk1md**
Compound name: **UNK2**
(...)
Compound name: **<CR>**
20 compounds loaded
Substructure list file: **substr.lis**
Results file:**<CR>**

Hits out of clauses

	UNK1	UNK2	UNK3	UNK4	UNK5	UNK6	UNK7	UNK8	UNK9
SS132	6/6	6/6	6/6	0/6	0/6	0/6	0/6	0/6	0/6
SS143	0/41	0/41	0/41	0/41	0/41	0/41	0/41	0/41	0/41

Hits out of clauses

	UNK10	UNK11	UNK12	UNK13	UNK14	UNK15	UNK16	UNK17	UNK18
SS132	0/6	2/6	0/6	0/6	2/6	0/6	0/6	0/6	0/6
SS143	0/41	0/41	0/41	0/41	0/41	0/41	0/41	15/41	4/41

Hits out of clauses

	UNK19	UNK20
SS132	0/6	0/6
SS143	3/41	0/41

subst	recall	reliability
SS132	100%	100%
SS143	100%	100%

FIGURE 6.9. Output generated by the RULE program for the analysis of 20 test compounds using rules for the phenothiazine and barbiturate substructures.

"RESULTS" file is then written which contains a list of substructures identified as present in an unknown for use by the automated structure generator in ACES.

The criterion for an identification to be made by the RULE program is the matching of at least one rule clause with spectral features in the MS/MS spectra of an unknown. One rule clause is sufficient to make an identification, because each rule clause has 100% reliability for the indicated substructure with respect to the reference database. The size of the reference database and the number of spectral features in a feature-combination affect the validity of this method of rule application. These considerations are addressed in Section 6.3.3.

6.3. EVALUATION OF MAPS SUBSTRUCTURE IDENTIFICATION RULES

Three different rule evaluations were performed on the MAPS rules generated from a reference database of 100 compounds. One of these rule evaluations used rule reliability and recall values, which are measures of rule accuracy and frequency of prediction. These equations reflect the reliability and recall of the rules *with respect to the reference database*. The reliability and recall obtained for a rule is dependent on the U_i and C_i program parameters used in the generation of the rule. However, given that the reference database contains only 100 compounds, these evaluative parameters do not *necessarily* reflect the reliability and recall of a rule when the rule is used on unknowns. Thus, the reliability and recall of the rules *with respect to the reference database* were used mainly in the optimization of the new MAPS program.

This discussion focuses on the remaining two evaluation procedures which test the accuracy of the MAPS rules. The first of these evaluations involved the comparison of the MS/MS spectral features contained in several MAPS rules with those contained in documented fragmentation pathways for the corresponding substructures. A large number of the compounds in the reference database were regulated drugs, since an important area of application of the MS/MS technique is in the analysis of pharmaceuticals. These analyses involve the screening of formulations for active drug components, impurities and synthetic markers, structural analyses of new drugs, and quantitation of drug metabolites in biological fluids.[1] Since pharmaceutically active drugs often have similar structures, MS/MS can be used to establish the structures of variants of more commonly encountered drugs.[1] MAPS rules for phenothiazine and barbiturate substructures are examined here in detail to illustrate the utility of the new MAPS software for generating reliable substructure identification rules for these compound classes. The remaining evaluation used 20 "test compounds" (i.e., compounds not in the reference database) to estimate the reliability and recall of the MAPS rules for identifying substructures in unknown compounds (see Section 6.3.3).

6.3.1. Evaluation of Rule Content

Several MAPS rules were examined to determine the correspondence of the spectral features found in the MAPS rules with those from documented fragmentation pathways.[20] These pathways have been discovered through the use of high-resolution mass spectrometry to determine the elemental composition of the various fragment ions and isotope labeling to assist in determining the mechanism of the fragmentation. Tandem mass spectrometry has also proven to be an effective tool in probing the ion chemistry of a variety of compounds.[1] The following sections provide a comparison of the spectral features found in MAPS rules for several substructures with those identified in the literature.

6.3.1.1. Phenothiazine

There are 13 compounds in the reference database which contain the phenothiazine substructure shown in Fig. 6.3. These compounds are clinically useful as antipsychotic drugs. An early study of phenothiazine derivatives used high-resolution mass spectrometry to determine the fragmentation pathways in preparation of the analysis of phenothiazine metabolites.[23] The investigators in this study divided the fragmentations of phenothiazine compounds into three groups: (1) fragments representing the side chain, (2) fragments representing the intact phenothiazine ring system with part of the side chain attached, and (3) fragments representing a partially fragmented ring system. MS/MS spectral features due to all three types of fragmentations have been observed in the MAPS rules for the phenothiazine substructure, depending on the U and C values used in generating the rules. The following discussion compares the spectral features contained in the MAPS rules for phenothiazine to those which were identified in the aforementioned study.

The initial features obtained for the phenothiazine substructure ($U_i = 40\%$ and $C_i = 70\%$) are shown in Table 6.1. No mass filter was utilized in this case. Thus, there are three spectral features (e.g., D 211.0) in this rule with masses larger than the nominal mass of phenothiazine (i.e., 198 amu). Structures of several fragment ions of phenothiazine and one substituent fragment ion are provided in Fig. 6.10. Pertinent fragmentation pathways for this substructure were gleaned from Refs. 10 and 12. These

TABLE 6.1. The Initial Features Obtained
for the Phenothiazine Substructure Using
the Indicated MAPS Program Parameters[a]

Feature	$[U, C]$	Feature #
D (211.0)	[40, 76]	{F1}
D (209.0)	[40, 76]	{F2}
D (198.0)	[41, 92]	{F3}
PD (198.0 → 171.0)	[76, 76]	{F4}
PD (198.0 → 154.0)	[92, 92]	{F5}
PD (197.0 → 196.0)	[68, 84]	{F6}
PD (197.0 → 153.0)	[90, 76]	{F7}
PD (196.0 → 152.0)	[90, 76]	{F8}
PD (70.0 → 27.0)	[45, 84]	{F9}

$(U_i = 40\%,\ C_i = 70\%,$ mass filter disabled$)$

[a]Reprinted with permission from Ref. 20.

pathways are listed in Table 6.2. We note that the phenothiazine "PD" features listed in Table 6.1 correspond directly to the key fragmentation pathways outlined in Table 6.2. The feature number from Table 6.1 is provided in brackets for each of the corresponding fragmentations shown in Table 6.2. The presence of the side chain ion in this list is the result of cross-correlation (see Section 6.3.2). These results demonstrate the ability of the MAPS rule generation software to select spectral features arising from the indicated substructure from within the MS/MS data space for use in generating substructure identification rules. The feature-combination rule shown in Fig. 6.3 was generated using many of the features shown in Table 6.1 except those that were removed using the mass filter. The reliability and recall of this rule were 100% with respect to the reference database.

TABLE 6.2. Comparison of Documented Fragmentation
Pathways for Phenothiazine Derivatives with the Features
Contained in the MAPS Phenothiazine Rule[a]

Fragmentation path $(m/z → m/z)$	Feature #	Corresponding neutral loss
198 (III) → 171 (IV)	{F4}	(loss of HCN—27 amu)
198 (III) → 154 (VII)	{F5}	(loss of CS—44 amu)
197 (IV) → 196 (V)	{F6}	(loss of H—1 amu)
197 (IV) → 153 (VIII)	{F7}	(loss of CS—44 amu)
196 (V) → 152 (IX)	{F8}	(loss of CS—44 amu)
70 (X) → 27 (XI)	{F9}	(loss of C_2H_5N—43 amu)

[a]Reprinted with permission from Ref. 20.

FIGURE 6.10. Ion structures identified for selected fragment ions of phenothiazine compounds[23] (reprinted with permission from Ref. 20).

6.3.1.2. Barbiturate

Barbiturates are another class of compounds that find use in pharmaceuticals and in the illicit drug trade.[24] There are 10 compounds which contain the barbiturate substructure in the reference database with a variety of substituents at the 1, 3, and 5 positions. The spectral features selected by the MAPS program for use in generating a barbiturate substructure identification rule are listed in Table 6.3. Values of 50% and 30% were used for initial uniqueness and correlation, respectively. The mass filter was disabled, so there are features in this list with masses greater than the nominal mass of the barbiturate substructure (128 amu). Ion structures for the parent ions found in Table 6.3 are provided in Fig. 6.11 except for

TABLE 6.3. Initial Features Obtained
for the Barbiturate Substructure
Using the Indicated Program Parameters[a]

Feature	$[U, C]$	Feature #
PD (125.0 43.0 P1)	[100, 30]	{F1}
PD (141.0 80.0 P1)	[100, 30]	{F2}
PD (98.0 80.0 P1)	[85, 46]	{F3}
PD (98.0 27.0 P1)	[83, 38]	{F4}
PD (169.0 97.0 P1)	[83, 38]	{F5}
PD (141.0 73.0 P1)	[80, 30]	{F6}
PD (167.0 124.0 P1)	[80, 30]	{F7}
PD (203.0 132.0 P1)	[80, 30]	{F8}
PD (124.0 43.0 P1)	[80, 30]	{F9}
PD (98.0 28.0 P1)	[75, 46]	{F10}
PD (169.0 126.0 P1)	[71, 38]	{F11}
PD (155.0 112.0 P1)	[66, 30]	{F12}
PD (189.0 118.0 P1)	[66, 30]	{F13}
PD (106.0 51.0 P1)	[66, 30]	{F14}
PD (112.0 66.0 P1)	[66, 30]	{F15}
PD (54.0 39.0 P1)	[66, 30]	{F16}
PD (160.0 133.0 P1)	[57, 30]	{F17}
PD (79.0 39.0 P1)	[57, 30]	{F18}
PD (112.0 94.0 P1)	[57, 30]	{F19}
PD (98.0 44.0 P1)	[55, 38]	{F20}
PD (169.0 57.0 P1)	[55, 38]	{F21}
PD (80.0 52.0 P1)	[50, 46]	{F22}
PD (141.0 98.0 P1)	[50, 30]	{F23}
PD (55.0 28.0 P1)	[50, 30]	{F24}
PD (54.0 27.0 P1)	[50, 46]	{F25}

$(U_i = 50\%, C_i = 30\%)$
(mass filter disabled, USORT enabled)

[a] Reprinted with permission from Ref. 20.

the m/z 124 ion, which is very similar in structure to m/z 125, and for m/z 106, which is unassigned. These structures were obtained from the literature and published fragmentation pathways for barbiturates.[24-26] It is noteworthy that these structures should not be considered as indicative of the structures of all ions in the reference database with the given m/z value due to the existence of isobaric ions in unit resolution mass spectra. For example, the m/z 203 barbiturate ions in the reference database appear to have at least 3 different ion structures (see Fig. 6.11) based on the structures of the barbiturate standards. The ions shown in Fig. 6.11 also demonstrate that the fragmentation of barbiturates are much more dependent on substituents than the phenothiazine compounds. But once again, it is apparent that the U/C selection criteria for MS/MS spectral features effectively limits the features used for rule generation to those that are directly related to the substructure of interest.

The process of generating feature-combinations during rule generation further limits the spectral features, because not all ions in the list of initial features are incorporated into the rules. For example, all of the feature-combinations generated for the barbiturate substructure with correlation

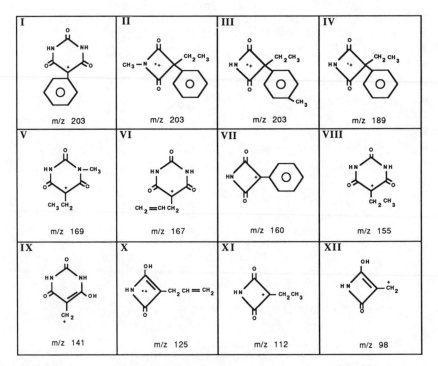

FIGURE 6.11. Ion structures identified for selected fragment ions of barbiturate compounds[24-26] (reprinted with permission from Ref. 20).

(C_c) greater than 37% involve daughter ions of m/z 98. These combinations are listed in Table 6.4. The m/z 98 parent ion, as shown by structure **XII** in Fig. 6.11, has lost most of the substituents that differentiate the barbiturate standards (e.g., allyl, ethyl, and phenyl groups). Therefore, the m/z 98 ion can be considered as the "lowest common denominator" among the barbiturate standards. One potential method for improving the barbiturate rules is to collect MS/MS data using an alternative ionization condition. It has been noted previously that the MAPS software does not limit the user to a particular set of operating conditions.[6] Rather, a reference database can be created using any set of MS/MS operating conditions (e.g., negative chemical ionization).

Feature-combinations with correlations lower than 37% (not listed in Table 6.4) include several of the ions shown in Fig. 6.11 with masses greater than 98 amu (i.e., structures **VI, VIII, X,** and **XI**). These combinations have lower correlation, because the features which comprise them are more specific to a particular subclass of barbiturates. For example, a feature-combination with a correlation of 23% (i.e., 3 out of 13 of the barbiturate standards) and involving m/z 203, m/z 189, and m/z 160 ions was obtained using the features listed in Table 6.3. All three of the barbiturate reference compounds which yield these MS/MS features were 5-phenyl barbiturates. Other feature-combinations also grouped a particular subclass of barbiturates (i.e., 5-ethyl and 5-allyl barbiturates). Once again, the feature-combinations involving m/z 98 were more generic in the barbiturate standards which contained this ion and therefore yielded a larger correlation value. This information is currently being explored to determine if more specific information can be provided to the structure generator of ACES when feature-combinations with specific "cross-correlations" are found in the MS/MS spectra of unknown compounds.

TABLE 6.4. The MAPS Feature-Combination Rule Obtained
for the Barbiturate Substructure[a]

Feature-combination (rule clause)	$[U, C]$
PD (98.0 28.0) and PD (98.0 55.0) and PD (98.0 70.0)	[100, 46]
PD (98.0 80.0) and PD (98.0 27.0) and PD (98.0 70.0)	[100, 38]
PD (98.0 80.0) and PD (98.0 27.0) and NL (111.0)	[100, 38]
PD (98.0 80.0) and PD (98.0 28.0) and PD (98.0 70.0)	[100, 38]
PD (54.0 27.0) and PD (98.0 55.0) and PD (98.0 70.0)	[100, 38]

Rule clauses with $C_c < 38\%$ not shown
REL = 100% and REC = 92% with respect to reference database
$U_i = 30\%$, $C_i = 30\%$, $C_c = 30\%$, MINF = 3, mass filter disabled

[a]Reprinted with permission from Ref. 20.

For example, if the barbiturate rule clause that had a high degree of cross-correlation with the phenyl substructure (i.e., all of the reference compounds associated with the feature-combination) had both the barbiturate and phenyl substructures, then the MAPS software could pass both of these substructures to the structure generator, possibly with a specific substitution pattern. The next section discusses cross-correlation which can affect the reliability of the MAPS rules for the analysis of compounds not in the reference database.

6.3.2. Cross-Correlation

Cross-correlation is a potential source of error in using MAPS due to the empirical nature of the MAPS algorithm.[20] This algorithm requires that any substructure be represented in the reference database in a number of different structural environments (i.e., with a variety of substituents). This requirement ensures that the software will be able to reliably associate the appearance of the common features within the MS/MS spectra of a set of compounds containing a specific substructure with only that substructure. If a large number of compounds in the reference database have two or more substructures in common and there are no other occurrences of these substructures in the database, the MAPS software will be unable to reliably assign spectral features to a particular substructure due to cross-correlation of the features. Thus, a reference database composed mostly of compounds with the same substructures will yield few reliable rules. Importantly, cross-correlations can be detected during the rule generation process and appropriate action can then be taken to remove, or at least reduce, the cross-correlation.

There are several methods for quantitatively assessing cross-correlation in the reference database.[20] One method is to calculate uniqueness values for a feature with respect to all defined substructures. This can be done on command in the MAPS software. If a feature has a high degree of uniqueness for more than one unrelated substructure, then the features are either cross-correlated or ambiguous. Features thus identified can be deleted from the feature list prior to rule generation. A more effective method for reducing the cross-correlation of features (in terms of maximizing the number of features available for rule generation) is to increase the variety of substructures represented in the reference database compounds.

A substructure cross-correlation coefficient can be calculated using Eq. (6.5) to determine if two substructures are cross-correlated.

$$XC(SS_j)_k = \frac{\text{number of compounds with } SS_j \text{ and } SS_k}{\text{number of compounds with } SS_j} \tag{6.5}$$

where SS_j and SS_k are substructures j and k, respectively.

Cross-correlation coefficients are calculated on command in the MAPS software for all defined substructures with respect to a given reference substructure. It is expected that a high degree of cross-correlation (i.e., 100%) will be observed for related substructures (i.e., substructures that are an integral part of other substructures). All other cross-correlation coefficients should be low for a "balanced" reference database. The remedy for cross-correlation, once again, is to add compounds to the reference database that have one, but not all, of the cross-correlated substructures.

6.3.3. Application of MAPS Rules to Twenty Test Compounds

Several MAPS rules for the phenothiazine and barbiturate substructures were applied to the MS/MS data of 20 test compounds to estimate the reliability and recall of these rules for unknown compounds.[20] The rules used in this test were generated using single collision, multiple collision, and both single and multiple collision reference data. In MS/MS experiments such as those performed on a TQMS, the pressure of the neutral collision gas (usually argon) can be set such that an ion selected by the first mass filter will likely encounter only a single target atom (i.e., single collision conditions) or more than one target atom (i.e., multiple collision conditions). The difference between these conditions is that the collision products obtained using single collision conditions are the result of the loss of a single neutral species, while the collision products obtained using multiple collision conditions can be the result of several successive neutral losses. This difference can have important implications for structure elucidation. Every MS/MS spectral feature involving a daughter ion is tagged (e.g., "P1") by the MAPS software to identify the collision conditions that were employed to acquire the pertinent daughter spectrum.

There are a number of different sets of MAPS program parameters which yield substructure identification rules with 100% reliability and recall with respect to the reference database. Briefly, optimal rule reliabilities are obtained using the current reference database with initial uniqueness and correlation values greater than 30%. Also, the highest possible MINF value (minimum number of spectral features per rule clause) should also be used. In theory, the presence of a single rule clause (a feature-combination) in the MS/MS spectra of an unknown should be sufficient for a substructure prediction to be made, since each rule clause has 100% uniqueness for the indicated substructure (with respect to the reference database). However, if a rule contains feature-combinations consisting of only two features, the substructure prediction may be based on the presence of only two spectral features in the MS/MS spectra of an unknown. Thus, the MINF parameter can be used to increase the number of spectral features which must be observed in the MS/MS spectra of an unknown before a substructure prediction is made.

The reliability and recall estimates (with respect to the reference database and with respect to the test compounds) obtained for the phenothiazine rules are listed in Table 6.5A. In this case, a MINF value of 2 and high values of U_i C_i were used to generate the rules. A high degree of recall (i.e., 100%) was obtained for the rules generated using the single collision and combined single and multiple collision data, while a recall of 76% was observed for the rule generated using only multiple collision data. Lower recall is often observed for rules generated using multiple collision data when high values for U_i and C_i and low values for MINF are used to generate the rules. A MAPS rule, not shown in Table 6.5A, was generated using lower U_i and C_i values (i.e., 30% and 30%, respectively) and an increased MINF value (i.e., 4). The recall of this rule was 100%. This rule also identified all five of the phenothiazine-containing compounds among the 20 test compounds (REL = 100% and REC = 100% w/r/t test compounds). This result tends to reinforce the idea that the rules should be optimized to include feature-combinations composed of several spectral features.

A reliability estimate of 100% was observed for the phenothiazine rules generated using single collision data and the combined single and multiple collision data. One false positive was obtained using the rule generated from the multiple collision data. This rule incorrectly identified the presence of the phenothiazine substructure based on one out of five rule clauses being true for a compound which did not have the phenothiazine substructure. All other predictions obtained using this rule were based on five out of five rule clauses being true for a test compound. The latter predictions provide greater confidence that a correct prediction has been made, since a larger number of spectral features were observed in the appropriate test compounds. Once again, the phenothiazine rule generated using the increased MINF value and multiple collision conditions correctly identified the 5 phenothiazine-containing test compounds with no false positives.

It was noted earlier that the fragmentation of barbiturates was much more dependent on side chains than the phenothiazine compounds. Consequently, it was more difficult to obtain a reliable rule for the barbiturate substructure (see Table 6.5B). One problem involved the smaller mass ions which incorporate only a fraction of the original barbiturate substructure but often have the highest correlation with the barbiturate substructure. These ions can be produced from compounds which are not barbiturates. The reference database appears to have been too small to prune out all of the false feature-combinations which included these features.

TABLE 6.5. Rule Reliability and Recall Values Calculated for (A) Phenothiazine Rules, (B) Barbiturate Rules Generated Using a Low MINF Value, and (C) Barbiturate Rules Generated Using a Higher MINF Value[a,b]

(A)	P1		P2		P1 + P2	
	REL	REC	REL	REC	REL	REC
w/r/t RDB	100	100	100	76	100	100
w/r/t TC	100	100	80	80	100	100
info	7/7 f.	6 cl.	8/8 f.	5 cl.	10/15 f.	7 cl.

$(U_i = 40\%, C_i = 70\%, C_c = 70\%$, mass filter on, MINF $= 2$, HITS $= 5$, MAXSF $= 10$, USORT selected)

(B)	P1		P2		P1 + P2	
	REL	REC	REL	REC	REL	REC
w/r/t RDB	100	92	100	84	100	76
w/r/t TC	50	100	60	100	60	100
info	30/56 f.	24 cl.	30/87 f.	26 cl.	30/129 f.	22 cl.

$(U_i = 30\%, C_i = 30\%, C_c = 30\%$, mass filter off, MINF $= 2$, HITS $= 5$, MAXSF $= 30$, USORT selected)

(C)	P1		P2		P1 + P2	
	REL	REC	REL	REC	REL	REC
w/r/t RDB	100	92	100	69	100	92
w/r/t TC	100	100	75	100	100	100
info	56/56 f.	20 cl.	87/87 f.	18 cl.	110/129 f.	20 cl.

$(U_i = 30\%, C_i = 30\%, C_c = 20\%$, mass filter off, MINF $= 4$, HITS $= 5$, MAXSF $= 110$, USORT selected)

[a]Reprinted with permission form Ref. 20.
[b]KEY: w/r/t, with respect to; RDB: reference database; TC: test compounds; P1: DB # 1 data—single collision conditions; P2: DB # 2 data—multiple collision conditions; P1 + P2: combined databases—each feature is tagged with P1 or P2; f.: # of initial features; cl.: # of rule clauses.

The results obtained for another set of barbiturate rules are summarized in Table 6.5C. We note that these rules were generated using a MINF value of 4. Once again, best rule application results are obtained using the single and the combined single and multiple collision data (REL = 100%, REC = 92% w/r/t reference database and REL = 100%, REC = 100% w/r/t test compounds). No false positives were observed using these rules. One false positive for the barbiturate substructure was observed for the rule generated with only the multiple collision data. Significantly, this identification was based on only a single rule clause being true. Also, the mis-identified compound was primidone, a compound containing a substructure very closely related to the barbiturate substructure (i.e., lacking only one carbonyl functionality). This compound was included in the test compounds to test the ability of the MAPS rules to discern closely-related substructures. The barbiturate rule generated using the combined data had no true clauses when applied to the MS/MS spectra of primidone.

Another method to increase the reliability of the MAPS rules for identifying substructures outside of the reference database is to increase the size of the reference database. A MAPS rule derived from a relatively large reference database (e.g., 1000 compounds) will have a greater inherent reliability than a rule derived from a smaller database (e.g., 100 compounds) because a greater number of false correlations will be eliminated from the rules. A larger reference database will also decrease the likelihood of cross-correlations, since more substructures will be found in a greater number of structural environments.

6.4. LIMITATIONS

The limitations of the current MAPS software include the lack of the "self-optimizing" algorithm and the range and purity of reference compounds which can be analyzed using current MS/MS instrumentation. In the first case, the MAPS code itself is limited because optimal program parameters vary among substructures. Thus, it is difficult to establish a set of parameters that can be used to generate a rulebase using the MAPS ALL command. Currently, rules must be generated and evaluated on an individual basis. In the second case, the acquisition time required to obtain a MS/MS map of the reference compounds limits the separation techniques that can be used to purify the reference standards.

An additional limitation involves the range of unknowns that can be analyzed by the ACES system. This limitation derives from the assumption which must be made by the automated structure generator that all substructure identifications made by the RULE program be assigned to one component. This assumption is valid for pure unknowns but not for the more often encountered mixture. If complete structure elucidation is not required, however, the MAPS rules can be used for substructure screening of mixtures. These limitations can be substantially reduced with further research. First, a set of heuristics can be added to the code to allow self-optimization. Several sets of program parameters may be tried using heuristics as a guide to provide optimal rules for all substructures. Second, development of a MS/MS instrument capable of acquiring MS/MS maps on the chromatographic time scale will assist in the acquisition of reference spectra and allow the ACES system to properly assign substructure predictions to the specific components of a mixture.

6.5. CONCLUSION

Computer-assisted interpretation of MS/MS spectra adds a significant new tool for structure elucidation of unknown organic compounds. Large portions of the structure of an unknown compound can be rapidly identified by application of substructure identification rules. The latest version of the MAPS software provides a comprehensive set of software tools for manipulating MS/MS and substructure data, the ability to generate substructure identification rules based on feature-combinations, and the ability to readily analyze the substructure content of an unknown using the RULE program. The use of an initial uniqueness and correlation value in the new MAPS program provides a convenient means of limiting the spectral features used for feature-combination rule generation to those features directly related to the indicated substructure. The application of rules for the phenothiazine and barbiturate substructures to 20 test compounds indicates that reliable substructure identification rules for unknowns can be generated using the MAPS software. Rules generated using combined single and multiple collision data add a new dimension (i.e., collision gas pressure dependence) to the information used by the MAPS software for structure elucidation.

While the current MAPS software is not "self-optimizing" so rulebases containing many rules are not easily generated, the addition of heuristics to the MAPS program which guide the selection of optimal program parameters should resolve this problem. Further expansion of the current reference database and creation of ancillary databases (e.g., using NCI so rules based on negative ions can be generated) will provide a robust expert system for structure elucidation using MS/MS spectra.

ACKNOWLEDGMENTS

The authors wish to thank Adrian Wade, Pete Palmer, Drake Diedrich, and Chris Weaver for their contributions to this project. This work is supported by National Institutes of Health grant GM-28254. Thanks are also due to Finnigan MAT and MSU for funds to purchase the VAXstation 3200 and to Molecular Design, LTD. for the source code to GENOA and STRCHK. KJH wishes to thank Oak Ridge Associated Universities and the Oak Ridge National Laboratory for support during the preparation of this manuscript.

REFERENCES

1. K. L. Busch, G. L. Glish, and S. A. McLuckey, *Mass Spectrometry/Mass Spectrometry: Techniques and Applications of Tandem Mass Spectrometry*, VCH Publishers, Inc., New York (1988).

2. F. W. McLafferty, *Tandem Mass Spectrometry*, Wiley, New York (1983).

3. A. L. Burlingame, D. Maltby, D. H. Russel, and P. T. Holland, "Fundamental Reviews—Mass Spectrometry," *Anal. Chem.* **60**, 300R–302R (1988).

4. B. D. Nourse and R. G. Cooks, "Aspects of Recent Developments in Ion-Trap Mass Spectrometry," *Anal. Chim. Acta* **228**, 1–21 (1990).

5. R. A. Yost and C. G. Enke, "Triple Quadrupole Mass Spectrometry for Direct Mixture Analysis and Structure Elucidation," *Anal. Chem.* **51**, 1251A–1264A (1979).

6. K. J. Hart and C. G. Enke, "An Automated Chemical Structure Elucidation System for MS/MS Spectra," *Chemometrics and Intelligent Laboratory Systems* (in press).

7. R. H. Johnson and U. Steiner, "Design Improvements in MS/MS Hardware and Software," *Pittsburgh Conference and Expositision Abstracts*, abstract # 541, Atlantic City, NJ (1986).

8. S. A. Lammbert, W. K. Chapman, U. Steiner, and A. E. Schoen, Data Dependent Instrument Control on a Triple Stage Quadrupole Mass Spectrometer," *Pittsburgh Conference and Expositision Abstracts*, abstract # 1105, Atlantic City, NJ (1987).

9. V. C. Parr, J. Waddicor, and D. Wood, "A Novel MS/MS Workstation," *Pittsburgh Conference and Expositision Abstracts*, abstract # 1523, Atlanta, GA (1989).

10. P. T. Palmer, *Algorithms for Interpretation of MS/MS Data for Chemical Structure Elucidation*, Ph.D. Dissertation, Michigan State University, East Lansing, MI (1988).

11. C. G. Enke, A. P. Wade, P. T. Palmer, and K. J. Hart, "Solving the MS/MS Puzzle: Strategies for Automated Structure Elucidation," *Anal. Chem.* **59**, 1363A–1371A (1987).

12. K. J. Hart, Ph.D. Thesis, Michigan State University (1989).

13. R. K. Linsay, G. B. Buchanan, E. A. Feigenbaum, and J. Lederburg, *Applications of Artificial Intelligence for Organic Chemistry—The Dendral Project*, McGraw-Hill, New York (1980).

14. A. P. Wade, P. T. Palmer, K. J. Hart, and C. G. Enke, "Development of Algorithms for Automated Elucidation of Spectral Feature/Substructure Relationships in Tandem Mass Spectrometry," *Anal. Chim. Acta* **215**, 169–186 (1988).

15. P. T. Palmer, K. J. Hart, and C. G. Enke, "Optimization of Automatically Generated Rules for Predicting the Presence and Absence of Substructures from MS and MS/MS Data," *Talanta* **36**, 107–116 (1989).

16. K. J. Hart, D. L. Diedrich, P. T. Palmer, and C. G. Enke, "Generation of Substructure Identification Rules using Feature-Combinations from MS/MS Spectra," *J. Am. Soc. Mass Spectrom.* (in press).

17. P. T. Palmer and C. G. Enke, "Programs for Molecular Formula Determination Employing Data from Unit Resolution MS/MS Spectra," *Int. J. Mass Spectrom. Ion Process.* **88**, 81–95 (1989).

18. M. H. Bozorgzadeh, "A Method for Determining the Elemental Composition of Daughter Ions in Simple Tandem Mass Spectrometry Spectra," *Rapid Commun. Mass Spectrom.* **2**, 61–64 (1988).

19. G. L. Glish, S. A. McLuckey, and K. G. Asano, "Determination of Daughter Ion Formulas by Multiple Stages of Mass Spectrometry," *J. Am. Soc. Mass Spectrom.* **1**, 166–173 (1990).

20. K. J. Hart, A. P. Wade, B. D. Nourse, and C. G. Enke, "Evaluation of Automatically Generated Substructure Identification Rules from MS/MS Spectra," *J. Am. Soc. Mass Spectrom.* (in press).

21. R. E. Carhart, D. H. Smith, N. A. B. Gray, J. G. Nourse, and C. Djerassi, "GENOA: A Computer Program for Structure Elucidation Utilizing Overlapping and Alternative Substructures," *J. Org. Chem.* **46**, 1708 (1981).

22. N. A. B. Gray, *Computer-Assisted Structure Elucidation*, Wiley, New York (1986).

23. J. N. T. Gilbert and B. J. Millard, "Pharmacologically Interesting Compounds—I: High Resolution Mass Spectra of Phenothiazines," *Org. Mass Spectrom.* **2**, 17–31 (1969).

24. F. C. Falkner and J. T. Watson, "Mass Spectrometry of the Trimethylsilyl Derivatives of Medicinal Barbiturates," *Org. Mass Spectrom.* **8**, 257–266 (1974).

25. J. T. Watson and F. C. Falkner, "The Mass Spectra of Some N-Substituted Barbitals and their Trimethylsilyl Derivatives," *Org. Mass Spectrom.* **7**, 1227–1234 (1973).

26. S. Dilli and D. N. Pillai, "The Synthesis of Volatile Barbituric Acid Derivatives," *Aust. J. Chem.* **28**, 2765–2774 (1975).

Canonical Correlation Analysis of Multisource Fossil Fuel Data

7

Henk L. C. Meuzelaar,
Miltiades Statheropoulos,
Huaying Huai, and Youngseung Yun

7.1. INTRODUCTION

7.1.1. Analysis of Multisource Fossil Fuel Data

Extremely complex materials, such as coals and other fossil fuels, pose special challenges for spectroscopists since most analytical problems involving such materials cannot be solved by applying any single spectroscopic method. Typically, mass spectrometry (MS) as well as infrared (IR) and nuclear magnetic resonance (NMR) spectroscopy data will need to be obtained when attempting to characterize the organic composition and structure of a fossil fuel sample. In addition, a host of other analytical tools, ranging from classical microscopic as well as elemental analysis

---

Henk L. C. Meuzelaar, Miltiades Statheropoulos, Huaying Huai, and Yongseung Yun • Center for Micro-Analysis and Reaction Chemistry, EMRC, University of Utah, Salt Lake City, Utah 84112.

Computer-Enhanced Analytical Spectroscopy, Volume 3, edited by Peter C. Jurs. Plenum Press, New York, 1992.

methods to specialized chromatographic and spectroscopic techniques, may have to be called upon.

Unfortunately, however, most members of the current generation of highly specialized spectroscopists are not familiar with more than one or two spectroscopic methods. Consequently, a researcher submitting fossil fuel samples or other complex materials to several different spectroscopy departments is likely to receive minimal assistance in trying to integrate and correlate the resulting multisource data. In our experience, canonical correlation analysis (CCA), a multivariate data analysis method,[1,2] is a powerful tool for exploring the relationships between two or more sets of spectroscopic data.[3-23]

Generally, each data set will consist of a so-called "two-way table of variance" in the form of a matrix with the rows representing the different sample spectra (objects) and the columns denoting the spectroscopic variables (features). In order to obtain interpretable CCA results, it is essential that the various matrices represent the same set of objects (and thus contain an equal number of rows) and/or the same suite of variables (equal number of columns).

For multisource spectroscopic data the spectroscopic variables will be different and thus each data set perferably should describe the same set of objects. Moreover, CCA software is generally designed to optimize the correlations between the data sets by rotating the object scores in the corresponding principal component spaces, as will be explained in the following pages. Unless the objects are identical in each data set, i.e., possess the same chemical relationships, the results of these rotations might not be readily interpretable.

Nonetheless, CCA may be quite informative in monitoring, or even modeling and predicting, chemical changes in a given set of samples, e.g., during a conversion process such as coal liquefaction.[14,16,21-23] In this case, however, the same spectroscopic technique is used, preferably under very similar experimental conditions. Consequently, the spectroscopic variables are identical in each data set and most variance can usually be attributed to the chemical process.

The following sections of this chapter will demonstrate the application of CCA to multisource fossil fuel data generated by the application of three different spectroscopic methods: Pyrolysis Low Voltage Electron Ionization Mass Spectrometry (LVMS); Pyrolysis Field Ionization Mass Spectrometry (FIMS); and Photo-acoustic Fourier Transform Infrared Spectroscopy (FTIR) to a single suite of eight U.S. coals obtained from the Argonne National Laboratory Premium Coal Sample Program (ANL-PCSP). Consequently, the three data sets obtained describe the same suite of samples but are markedly different with respect to number as well as type of variables.

7.1.2. Objectives and Principles of Canonical Correlation Analysis

One of the main objectives in applying CCA to analytical spectroscopy is extract chemical information from the study of the relationships among two or more sets of chemical data.[3-23] This is achieved by deriving for each set a weighted linear combination of the original variables in such a way that the two linear combinations (canonical variates) are maximally correlated. Subsequently, additional linear combinations, each of which is independent of the preceding pair(s) of linear combinations, are derived that maximize the remaining correlations. Often CCA is carried out on the most significant principal components of each data set. In this case, CCA may be regarded as an orthogonal rotation of the principal component axes in such a way that lower-dimensional subspaces are found for all data sets in which the relative positions (coordinates) of the sample spectra are most highly correlated.

Consequently, the numerical output of a typical CCA procedure consists of the "loadings" (correlation coefficients) of the original variables and/or the intermediate principal components on each of the Canonical Variate (CV) functions, the "scores" (coordinates) of the objects in CV space, the "canonical correlations" (R), and various measures of the statistical significance of each CV function.

7.1.3. Mathematical Rationalization of CCA

The general mathematical problem in CCA, as originally described by Hotelling,[24,25] is to establish the maximum correlation among sets of variables. For two sets of variables the mathematical solution can be extracted by first performing independent principal component analysis (a type of factor analysis) on each of the two sets of variables. In this way multi-collinearities are removed from the variable sets prior to CCA. Then the resulting component vector axes are rotated to develop weights for each variable that produce maximal correlations between components on each side.

Assuming that the standardized data for two sets of variables are expressed by matrices X_m and X_w (sizes $c \times m$ and $c \times \lambda$, respectively), then the corresponding correlation matrices V_m and V_w can be calculated as follows:

$$V_m = (1/c) X_m^T X_m \qquad (7.1)$$

and

$$V_w = (1/c) X_w^T X_w \qquad (7.2)$$

Here, c denotes the number of spectra and m, λ denote the number of variables. Subsequently, rotation matrices E_m and E_w must be found that maximize the variance such that

$$\Lambda_1 = (1/c)(X_m E_m)^T (X_m E_m) \tag{7.3}$$

and

$$\Lambda_2 = (1/c)(X_w E_w)^T (X_w E_w) \tag{7.4}$$

which can be written in the alternative form

$$\Lambda_1 = E_m^T V_m E_m \tag{7.5}$$

and

$$\Lambda_2 = E_w^T V_w E_w \tag{7.6}$$

The Λ matrices are diagonal with the diagonal elements ranked in descending order. This is a mathematical problem of the eigenvector–eigenvalue type. The factors can be calculated in the following way:

$$F_m^T = \Lambda_1^{1/2} E_m^T \tag{7.7}$$

and

$$F_w^T = \Lambda_2^{1/2} E_w^T \tag{7.8}$$

In this way the original data can be reduced to the score matrices

$$S_m = X_m (E_m \Lambda_1^{-1/2}) \tag{7.9}$$

and

$$S_w = X_w (E_w \Lambda_2^{-1/2}) \tag{7.10}$$

Then, for these score matrices S_m and S_w the corresponding correlation matrix

$$G_{mw} = (1/c) S_m^T S_w \tag{7.11}$$

is calculated. Subsequently, the canonical correlation problem is defined in finding two matrices T_m and T_w so that $S_m T_m$ and $S_w T_w$ correlate maximally. The diagonal matrix

$$MW = \frac{(S_m T_m)^T (S_w T_w)}{c} = T_m^T G_{mw} T_w \tag{7.12}$$

is calculated. The solutions for T_m and T_w can be determined with standard mathematical procedures.

For more than two sets of variables various solutions have been proposed.[26,27] One of the procedures[27] defines and solves the problem as follows: for each of n sets of variables the canonical correlation problem is to find a single canonical variate consisting of a suitable linear combination of the variables that maximizes the sum of the weighted squared correlations of these (n) linear composites with the single $(n+L)$th variate X_i representing k observations on M_i variables. Thus, we seek a k-vector Z (a vector Z with k elements) and n transformation vectors A_i (A_i has m_i components) such that

$$R^2 = \sum_{i=1}^{n} w_i [r(Z, A_i X_i)]^2 \tag{7.13}$$

is maximum, where $r(Z, A_i X_i)$ denotes the correlation matrices of the Z and the linear composites $A_i X_i$. The quantity w_i is a weighting factor for the ith variable set. Now

$$\max[r(Z, A_i X_i)]^2 = [ZX_i^T (X_i X_i^T)^{-1} X_i Z^T]/(Z^T) \tag{7.14}$$

therefore

$$R^2 = ZQZ^T/ZZ^T \tag{7.15}$$

where

$$Q = \sum_{i=1}^{n} w_i X_i^T (X_i X^T)^{-1} X_i \tag{7.16}$$

For maximizing R^2 the eigenvector of Q corresponding to the largest eigenvalue has to be determined. This is a standard mathematical problem. The second best canonical Z variate orthogonal to the first is given by the second eigenvector of Q.

Besides the pure mathematical solution of the canonical correlation problem, the interpretation of the solutions as well as their statistical significance is an important aspect. Various tests and indices have been developed for determining the statistical significance as well as the nature of the canonical relationships.[28-30] Here we will focus primarily on techniques and procedures for interpreting the information contained in the CV subspaces.

7.2. EXPERIMENTAL

7.2.1. Sample Selection

The eight coals which we used for this study represent the entire Argonne National Laboratory Premium Coal Sample Program collection and include various ranks (= degrees of coalification) as well as maceral (microscopically recognizable organic components) composition (see Table 7.1.). Details of coal sample preparation and characterization procedures can be found elsewhere.[31]

7.2.2. Curie-Point Pyrolysis-Low Voltage Electron Ionization MS (LVMS)

Py-LVMS experiments were performed using an Extranuclear 5000-1 Curie-point Py-MS system with specially designed inlet as described else-

TABLE 7.1. Origin, Rank, Elemental Composition,[a] and Ash Content of the Eight Coal Samples

---

1. Beulah-Zap Seam (North Dakota) Lignite
   (C = 73, H = 5.3, O = 21, S = 0.8, ash = 6)

2. Wyodak-Anderson Seam (Wyoming) Subbituminous
   (C = 74, H = 5.1, O = 19, S = 0.5, ash = 8)

3. Illinois #6 Seam (Illinois) High Volatile Bituminous
   (C = 77, H = 5.7, O = 10, S = 5.4, ash = 16)

4. Blind Canyon Seam (Utah) High Volatile Bituminous
   (C = 79, H = 6.0, O = 13, S = 0.5, ash = 5)

5. Pittsburgh #8 Seam (Pennsylvania) High Volatile Bituminous

6. Lewiston-Stockton Seam (West Virgina) High Volatile Bituminous
   (C = 81, H = 5.5, O = 11, S = 0.6, ash = 20)

7. Upper Freeport Seam (Pennsylvania) Medium Volatile Bituminous
   (C = 87, H = 5.5, O = 4, S = 2.8, ash = 13)

8. Pocahontas #3 Seam (Virginia) Low Volatile Bituminous
   (C = 91, H = 4.7, O = 3, S = 0.9, ash = 5)

---

[a] % dry, ash free coal.

where.[32,33] The coal sample was hand-ground into a fine, uniform suspension in Spectrograde methanol (5 mg of sample per ml of methanol). Average particle size was approximately 5–20 μm. Five μl of the suspension were coated on the 0.5-mm-diameter ferromagnetic wire (Curie-point temperature 610 °C) and air-dried under continuous rotation for approximately 1 min, resulting in the deposition of approximately 25 μg of dry coal sample on the wire. The coated wire was inserted into a borosilicate glass reaction tube, introduced into the vacuum system of the mass spectrometer, and inductively heated at a rate of approximately 100 K/s to the Curie-point temperature (610 °C). MS conditions were as follows: electron impact energy 12 eV, mass range scanned 50–200 amu, total number of scans 41, total scan time 15 s. Total ion counts per spectrum were normalized to 100% in order to correct for differences in sample amounts.

7.2.3. Pyrolysis-Field Ionization MS (FIMS)

The experimental setup of the Finnigan MAT field ionization MS system has been described elsewhere.[34,35] About 100 μg of coal were placed in a quartz crucible, which was inserted into the high vacuum (10^{-7} torr) of the ion source using an AMD Intectra direct introduction system (Beckeln, FRG).[35] The crucible was heated from 50 to 730 °C at a linear heating rate of 100 K/min. The oven temperature was measured at the bottom of the crucible using a thermocouple. Moreover, the ion source was maintained at a temperature of 200 °C. The tip of the AMD probe, the wall of the ionization chamber, and the FI emitter were held at the same electric potential of 8 kV. Emitters were prepared from 100 μm tungsten wire by activation benzonitrile at 1200 °C.[36] Thirty-four mass spectra were recorded for Pittsburgh #8 coal in the m/z 50–900 mass range. Between repetitive scans the emitter was flash-heated to approximately 1500 °C in order to avoid condensation and to regenerate the surface of the emitter,[35] thereby increasing overall reproducibility of the Py-FIMS profiles.

7.2.4. Photoacoustic Fourier Transform Infrared Spectroscopy (FTIR)

Approximately 10 mg aliquots of ~100 mesh coal samples were used without diluent. Analyses were performed by means of a Perkin-Elmer model 1760-X FTIR spectrometer with MTEC Model 200 photoacoustic detector and small sample cup (<0.25 cm^3). The spectra were recorded in the 4000–500 cm^{-1} range using a spectral resolution of 4 cm^{-1}, minimum mirror velocity 0.1, and 128 scans. The sample chamber was purged with helium in order to remove carbon dioxide and water vapor.

7.2.5. Multivariate Data Analysis

Computerized multivariate data analysis was carried out using the SIGMA program package developed at the University of Utah Center for Micro Analysis and Reaction Chemistry[37] and involved Principal Component, Discriminant Rotation, and Canonical Correlation Analysis. After standardization of the raw data, Principal Component Analysis (correlation around the mean) was used as a data reduction method. The principal component scores were subjected to Discriminant Rotation and Canonical Correlation Analysis. Discriminant rotation is applied when replicate samples are available and attempts to maximize the ratio of "between-category" variance and "within-category" variance (variance between replicates). The CCA technique used is based on the generalization to three or more sets of variables described by Carroll.[27] The algorithms used for manipulating and solving matrices were published by Wilkinson and Reinsch.[38] Subsequently, the data were expressed as chemical components by using the VARDIA[15] technique. The chemical components extracted are presented in the form of bar plots, called "factor spectra."[39] (The terms principal component and factor are used interchangeably throughout the text.) This is a way of transforming the loadings into a form comparable to the original spectra. This transformation is done by multiplication of the loadings with the standard deviation of the mass variable involved. The "component axes" are directions in factor space and are attributed to specific chemical components. CCA determines directions in the factor spaces in which the scores of the different data sets have a maximum correlation.

7.3. MULTISOURCE DATA INTERPRETATION

7.3.1. Visual Inspection of the Original Data Sets

Spectra of three ANL coals obtained by LVMS, FIMS, and FTIR are shown in Figs. 7.1, 7.2, 7.3, respectively. The three coals represent different "ranks" (degrees of coalification) ranging from lignite (Beulah Zap seam) through high volatile bituminous (Blind Canyon seam) to low volatile bituminous (Pocahontas seam). The LVMS profile of the low-rank Beulah Zap coal in Fig. 7.1 is dominated by oxygen-substituted alkylaromatic signals, e.g., alkylphenols and alkyldihydroxybenzenes, while the high-rank Pocahontas coal produces a spectrum with prominent alkylaromatic hydrocarbon series, e.g., alkylbenzenes, alkylnaphthalenes, and alkyl-substituted polynuclear condensed aromatics.

At first sight, the Blind Canyon coal spectrum in Fig. 7.1b presents

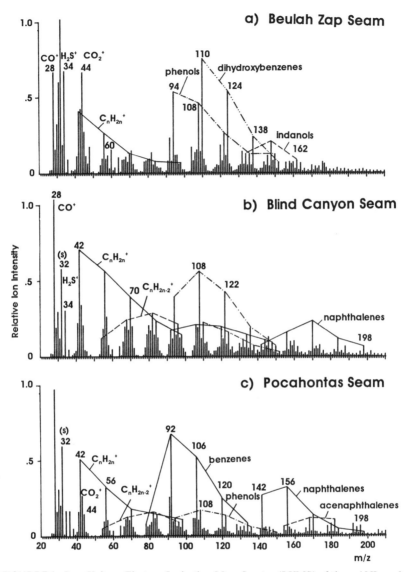

FIGURE 7.1. Low Voltage Electron Ionization Mass Spectra (LVMS) of three ANL coals.

a pattern of aromatic signal intensities commensurate with its intermediate rank. Upon closer examination, however, we note the relatively high abundance of aliphatic hydrocarbon signals at m/z intensities corresponding to the generalized formulas $C_nH_{2n}^+$ (e.g., alkenes) and $C_nH_{2n-2}^+$ (e.g., alkadienes). This reflects the unusually high resinite content (11%) of the Blind Canyon coal and thus is not directly related to rank but to the

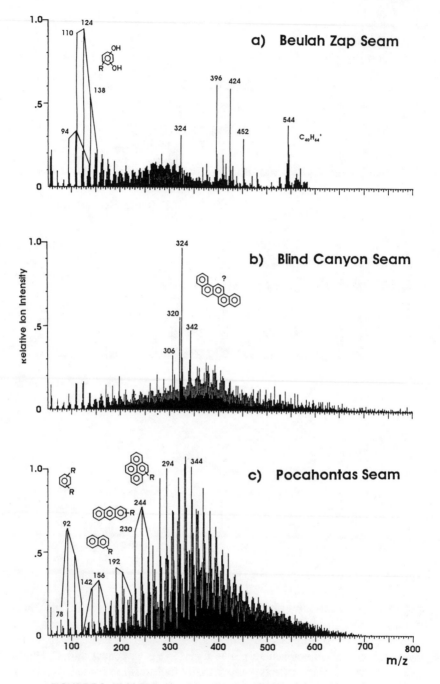

FIGURE 7.2. Field Ionization Mass Spectra (FIMS) of three ANL coals.

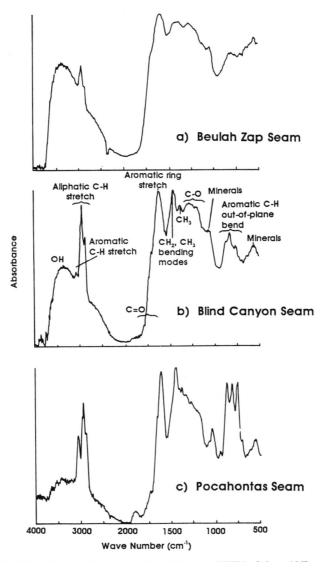

FIGURE 7.3. Fourier Transform Infrared Spectra (FTIR) of three ANL coals.

type of paleoenvironment in which the Blind Canyon coal was deposited some 70 million years ago.

Biomarker (chemical fossil) signals, which can be directly related to depositional environment, are clearly observable in the FIMS spectrum of Beulah Zap coal in Fig. 7.2a at m/z 396, 424, 452 (mass peaks believed to represent wax esters from plant leaf cuticles) and at m/z 544 (β-carotene).

The Blind Canyon coal spectrum in Fig. 7.2b is also dominated by several biomarker signals. Previous work has established a direct link with the high fossil resin content of this coal. In fact, most of these biomarker signals represent more or less strongly aromatized, resin-derived triterpenoids, e.g., in the form of 5 ring aromatics known as picenes.[40]

None of the higher MW biomarker signals observed by FIMS is seen in the LVMS profiles in Fig. 7.1, since the mass range covered does not extend above m/z 240. However, the low mass portions of Fig. 7.1 and Fig. 7.2 show a definite degree of correspondence for the Beulah Zap and Pocahontas coals. On the other hand, the strong aliphatic compound series seen in the lower mass range of the LVMS profile of Blind Canyon coal (Fig. 7.1b) is not reflected in the corresponding FIMS profile (Fig. 7.2b). In other words, there are marked differences as well as similarities between the LVMS and FIMS profiles. Obviously, these differences and similarities would be difficult to evaluate and/or quantitate in an objective manner without the use of multivariate statistical methods capable of handling spectra with hundreds of variables.

This is even more true when trying to compare the Py-MS data with the corresponding FTIR spectra in Fig. 7.3. Some general trends observed in Fig. 7.1 and 7.2, such as an increase in aromatic hydrocarbon signals accompanied by decreasing hydroxyaromatic signals in the higher-rank coals as well as the relatively high aliphatic hydrocarbon content of Blind Canyon coal (Fig. 7.3b), are readily confirmed by visual inspection of the FTIR spectra in Fig. 7.3. On the other hand, mineral matter signals, largely reflecting differences between depositional environments, are known to contribute significantly to the FTIR spectra of coal, e.g., in the 4000–2700 and 1800–600 cm^{-1} wavelength region, while neither of the two MS techniques used is likely to detect much of the mineral matter components of coal. Furthermore, major differences in responsivity toward different structural moieties complicate an attempt at direct visual comparison between the Py-MS and FTIR data. This is illustrated by the prominent 1800–1650 cm^{-1} FTIR region, known to represent carbonyl and related oxygen functional moieties, which have no equally prominent counterpart in the Py-MS spectra.

In order to investigate the usefulness of canonical correlation analysis for comparing the information present in the three coal data sets, it is desirable to first identify the major underlying chemical trends in each of the individual sets and to establish the respective intrinsic dimensionalities. It has been shown by Windig et al.[15] that this can be done by factor analysis (principal component analysis) followed by suitable factor rotation methods, e.g., using the so-called Variance Diagram (VARDIA) technique. Furthermore, transforming the original data to a lower-dimensional, orthogonal data spaces, such as one achievable by principal component analysis, has the added advantage of making the subsequent canonical correlation analysis manipulations more transparent, since the latter can now be simply regarded as orthogonal rotations of principal component axes aimed at maximizing the correlations between the data sets.

7.3.2. Principal Component Analysis of the LVMS Data

Principal component analysis of the LVMS data set consisting of 32 spectra (8 coals analyzed in quadruplicate) produced approximately 7 significant principal components, as judged from the scree plot in Fig. 7.4 using a "tail break" criterion. Score plots of the first four factors, together explaining 84% of the total variance in the data set, are shown in Fig. 7.5a and b. Considering the fact that the eight ANL coals were numbered in order of increasing rank, a strong rank trend can be seen in the F1/F2 space shown in Fig. 7.5a. The F3/F4 score plot (Fig. 7.5b) is dominated by two outliers: coals no. 3 (a high sulfur Illinois #6 coal) and no. 4 (a resin-rich Blind Canyon coal), and thus appears to represent primarily differences between ancient depositional environments.

Another interesting observation is that the within-category variance ("nonreproducibility") is not randomly distributed among the quadruplicates in F1–F4 space. Rather, most of the within-category variance appears to load on F2 and on a combined F3/F4 vector. Therefore, we can

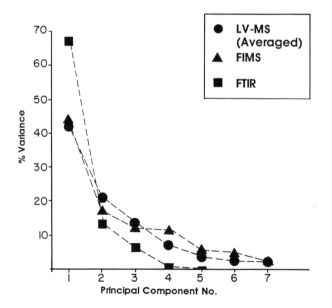

FIGURE 7.4. Scree plot of the principal components of three data sets.

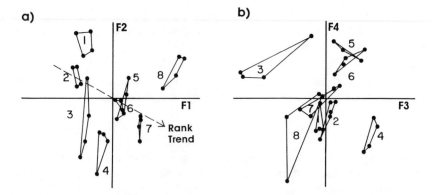

FIGURE 7.5. Factor score plots of LVMS data in the factor F1/F2 (a) and F3/F4 (b) spaces.

attempt to find directions in which between-category/within-category ratios are maximized, thereby effectively filtering out much of the within-category variance. The results of this so-called "discriminant rotation" approach are illustrated in Figs. 7.6a and b. In spite of the markedly improved separation between the eight coal categories, and/or the decreased variation between quadruplicate analysis within each category, the new discriminant score plots do not show a clear rank trend. Apparently, the greatest level of discrimination between the eight coal samples is not so much achieved by differences in rank as by other factors, such as characteristic differences in depositional environments. This is not surprising in light of the philosophy behind the ANL-PCSP coal collection which attempts to present the greatest possible level of diversity among a strictly limited number of economically important U.S. coals.

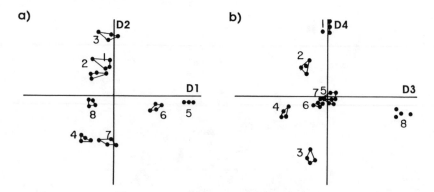

FIGURE 7.6. Plot of scores of discriminant functions D1/D2(a) and D3/D4 (b) obtained from LVMS data.

This explains the relatively high intrinsic dimensionality of the data sets. As a rule of thumb, the number of independent observations should be 2–3 times the intrinsic dimensionality of the data set before supervised manipulations of the data space are allowable. As seen in Fig. 7.4, the intrinsic dimensionality appears to be ≤ 5. The number of independent observations regarding the distribution of the between-category variance is only 8 while we have as many as 32 independent observations regarding the distribution of the total (between-category plus within-category variance). Consequently, it would appear that we have plenty of observations to allow examination and manipulation of the data space with regard to total variance or within-category variance related issues, but that extreme caution is necessary when trying to evaluate and interpret the distribution of the between-category variance.

At this point we need to decide whether to use the discriminant rotation scores in Fig. 7.6 or the original factor scores in Fig. 7.5 as the basis for subsequent CCA with the other two data sets. Although it may seem attractive to opt for the discriminant score data, it should be remembered that the discriminant space shown in Fig. 7.6 simply represents a four-dimensional subspace of the original seven-dimensional principal component space. In other words, depending on the selected cutoff level of the discriminant functions we effectively are discarding the factor space dimensions dominated by within-category, i.e., experimentally less desirable, sources of variance. Unfortunately, selection of appropriate cutoff levels for discriminant functions is even more of an art than the well-known dilemma of selecting cutoff levels for principal components. The main concern stems from the consideration that part of the relevant chemical information may well be imbedded in the within-category variance, and *vice versa*. This is especially worrisome when attempting to correlate data sets with potentially low levels of common variance, e.g., MS and FTIR data. Rigorously discarding the within-category variance, even though each of our categories consists only of replicate analyses, could result in the loss of precious information.

Moreover, it should be remembered that truly undesirable sources of variance, e.g., those associated with nonreproducibilities of sample preparation and/or instrumental analysis, are unlikely to correlate strongly when comparing two or more data sets obtained by different spectroscopic techniques. In other words, a substantial degree of filtering of unwanted sources of variance may be expected as a bonus when performing CCA.

For the above reasons, we often prefer to use original principal components for canonical correlation analysis purposes, although care must be taken to keep the total number of components selected low enough to avoid overdimensioning the data space. Often, the number of principal components which need to be retained can be determined only by trial and error, while monitoring valuable indicators of the variance partitioning process taking place during canonical correlation analysis, e.g., Wilk's coefficients, chi-square coefficients, degrees of freedom, etc. For an in-depth discussion of such indicators the reader is referred to the specialized literature, e.g., Refs. 1 and 28. Based on the above considerations and approaches, we decided to use the first seven principal components. Moreover, because canonical correlation analysis cannot be performed on categorized data, each category of four replicate analyses was reduced to a single data vector represented by its centroid (Fig. 7.7).

Before proceeding with correlations between the LVMS principal component scores and those of the other two data sets, however, careful inspection of the principal component loadings is recommended in order to obtain an overall impression of the chemical information present in the main dimensions of the data space. As shown by Windig et al.,[15] the Variance Diagram (VARDIA) method offers a convenient way of examining the factor loadings of mass spectral data and often facilitates the location of major chemical components and trends in factor space, provided that factor scores have been normalized in such a way that the cosine of the angle ϕ between any two original mass variable vectors χ_p and χ_q directly corresponds to the correlation coefficient of χ_p and χ_q. In other words, perfectly correlated mass variable vectors should have angles ϕ or 0 or 180 degrees (cosine ϕ = correlation coefficient χ_p, χ_q = $+1.0$ or -1.0, respectively), while noncorrelated variable vectors are orthogonal to each other in factor space and thus have angles ϕ of 90 or 270 degrees (cosine ϕ = correlation coefficient χ_p, χ_q = 0). Under these conditions, highly correlated mass variables tend to occur in densely clustered bundles. This behavior is highlighted by the VARDIA method, which sums the total variance found within narrow (e.g., 10°) sectors of two-dimensional principal component space, e.g., corresponding to the F1/F2 and F3/F4 planes, as illustrated in Figs. 7.8a and b, respectively.

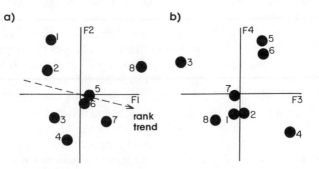

FIGURE 7.7. Factor score plots of LVMS data in the factor F1/F2 (a) and F3/F4 (b) spaces using the centroids of quadruplicate analysis.

The main "rank trend" discernible in F1/F2 score space (Fig. 7.7a) corresponds to a major lobe of correlated variables (labeled A) in Fig. 7.8a, apparently associated with the two lowest-rank coals (nos. 1 and 2 in Fig. 7.7a). Further inspection of the variables loading on component A is possible by plotting the so-called "factor spectrum" shown in Fig. 7.9a, using the procedure described by Windig *et al.*[39] Factor spectrum plots are produced by multiplying the loading of each variable on a given factor by the standard deviation of the variable, i.e., in effect plotting the covariance. This has the advantage of preserving some information regarding the original intensities of the mass variables, thereby producing mass spectral patterns that can be more readily compared to the original mass spectra. This is illustrated by comparing the numerically extracted "low rank" component pattern in Fig. 7.9a with the mass spectrum of the low-rank Beulah Zap coal in Fig. 7.1a. Similarly, component B in Fig. 7.9b has many features in common with the mass spectrum of the highest-rank coal (Pocahontas seam) in Fig. 7.1c.

Also, component C (Fig. 7.8a) can be seen to correspond with the direction of coal no. 4 (Blind Canyon seam) in the Fig. 7.7a score plot. Consequently, the most characteristic features of the mass spectrum of Blind Canyon coal in Fig. 7.1b are reflected in the numerically extracted pattern of component C in Fig. 7.9c. Finally, Fig. 7.9d shows a mass spectral component (F) with prominent sulfur-containing ion signals. By comparing Figs. 7.7b and 8b, the direction of component F can be seen to correspond with coal no. 3 (Illinois #6 seam), which is indeed known to contain more organic and inorganic sulfur compounds than any of the other seven coals.

7.3.3. Principal Component Analysis of the FIMS Data

Similar data-analysis strategies are followed when analyzing the Py-FIMS data. Although triplicate FIMS runs were available for all eight

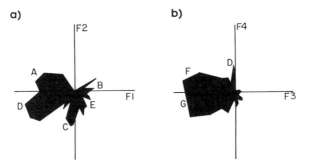

FIGURE 7.8. Variance diagrams showing the presence of six major components in the space spanned by factors F1/F2 (a) and F3/F4 (b), respectively (LVMS data).

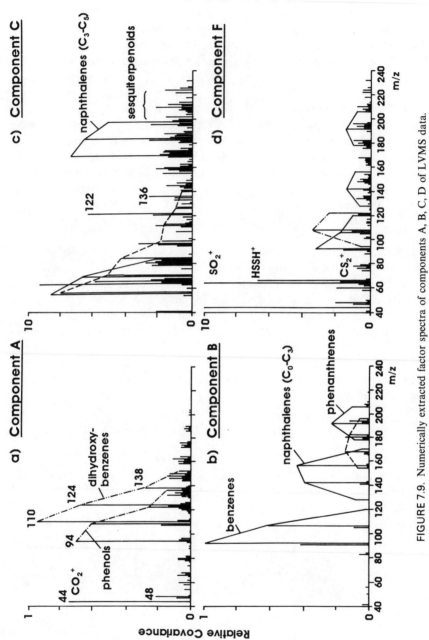

FIGURE 7.9. Numerically extracted factor spectra of components A, B, C, D of LVMS data.

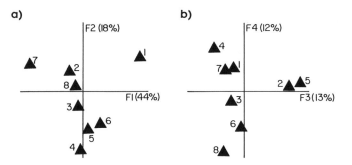

FIGURE 7.10. Factor score plots of FIMS data in the factor F1/F2(a) and F3/F4(b) spaces.

coals, the large number of FIMS variables dictated the use of averaged spectra (centroids) for each category in order to maintain acceptable computer-response times. Principal component analysis of the 8×750 matrix took approximately 30 minutes. As shown in Fig. 7.4, the dimensionality of the FIMS principal component space is relatively high with at least four, and possibly six, significant components. This would appear to be due to the presence of prominent higher mass range signals representing biomarker compounds and other bitumen-like components, which tend to be highly characteristic for each of the coals analyzed. The F1/F2 and F3/F4 score plots in Fig. 7.10 fail to show a clear rank trend. Rather, each dimension appears to be dominated by one or two characteristic spectra only. This is confirmed by the VARDIA plots in Fig. 7.11. Nonetheless, numerical extraction of the major correlating trends and components (see Fig. 7.12) reveals some clear similarities between the FIMS and LVMS data. For instance, component A in Fig. 7.12a, corresponding to the low-rank Beulah Zap lignite, shows similar prominent ion series at m/z 110/124/138 (dihydroxybenzenes and/or methoxyphenols) as observed in

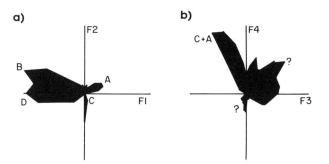

FIGURE 7.11. Variance diagrams showing the presence of various components in the space spanned by factors F1/F2 (a) and F3/F4 (b), respectively (FIMS data).

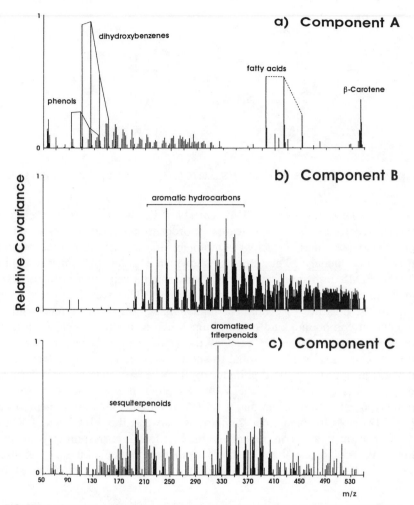

FIGURE 7.12. Numerically extracted factor spectra of components A, B, C of FIMS data.

Fig. 7.9a, in spite of the obvious differences in mass range and ionization methods. Further similarities can be observed in components B and C when compared with the corresponding LVMS components in Figs. 7.9b and c, respectively. Nevertheless, the relative orientation of the eight objects and the corresponding chemical components in the F1/F2 and F3/F4 projections of the FIMS F1–F4 space is quite different from those in LVMS space, as can be appreciated by comparing Figs. 7.7 and 7.8 with Figs. 7.10 and 7.11, respectively. Apparently, the principal component spaces of the LVMS and FIMS data sets are closely related chemically but, due to the different responses of the two techniques to similar chemical components

TABLE 7.2. Correlation between LVMS and FIMS Principal Components

FIMS	Py-LVEIMS				
	F1	F2	F3	F4	F5
F1	−0.375	0.587	0.117	0.005	−0.275
F2	−0.067	0.636	−0.436	−0.305	0.586
F3	−0.211	0.278	0.165	0.438	0.047
F4	−0.558	−0.300	0.440	−0.277	0.403
F5	−0.103	−0.234	−0.394	0.790	0.395

and trends, the relative amount of variance explained by the respective components varies considerably between the two data sets. As a result, the two principal component spaces may be regarded as rotated relative to each other. In view of the fact that CCA (as used here) can be regarded as an orthogonal rotation of principal component space, the LVMS and FIMS data sets would seem to be good candidates for subsequent CCA.

7.3.4. Canonical Correlation of the LVMS and FIMS Data

The strongly rotated nature of the chemical components distributed over both principal component spaces is illustrated in the matrix of cross correlations between the two principal component spaces is illustrated in the matrix of cross correlations between the two principal component spaces shown in Table 7.2. The strongest (correlation coefficient $\pm = 0.79$) is found between LVMS F4 and FIMS F5, followed by F2 vs. F2 ($\pm = 0.64$) and F5 vs. F2 ($\pm = 0.59$).

As illustrated in Table 7.3, CCA can conceptually be regarded as an attempt to diagonalize the matrix of cross correlations. Due to the high degree of correspondence between the two data sets (both methods are based on vacuum pyrolysis in front of the ion source of a mass spectrometer operating in a soft ionization mode) unusually high correlation coefficients

TABLE 7.3. Correlation between LVMS and FIMS CV Functions

FIMS	LVMS				
	CV1	CV2	CV3	CV4	CV5
CV1	1.000	0.000	0.033	−0.001	0.001
CV2	0.001	1.000	0.001	0.002	0.001
CV3	0.003	−0.001	1.000	0.000	−0.001
CV4	−0.002	0.002	0.001	1.000	0.002
CV5	0.000	−0.001	0.002	−0.001	1.000

are found. The 1.00 scores in the diagonal of Table 7.3 constitute a low probability, near-perfect outcome of the canonical correlation procedure and may be the result of overdimensioning the data space. The important point here is that even the most experienced data analyst would be hard pressed to predict Table 7.3 from a visual evaluation of Table 7.2 alone. The highly successful outcome of the canonical correlation rotation of the two data spaces is further illustrated in the CV1/CV2 score plot in Fig. 7.13. In view of the limited number ($n = 8$) of observations, it is conceivable that high correlations might be obtained through chance alone. A first-line defense against chance correlations is to examine the amount of shared variance. In this case the variance found in CV1/CV2 space represents 38.6% of the total variance. This is a respectable quantity and again confirms the highly overlapping nature of the chemical information obtained by both techniques.

Once the CV subspaces have been defined for each data set, we may proceed with the extraction of chemical components as a second method of checking the chemical relevance of the rotations obtained. Again, the VARDIA method can help in locating major components and trends in CV space. When applied to our data set, this approach (not shown here) reveals the same chemical components already observed in the respective principal component spaces.

7.3.5. Principal Component Analysis of the FTIR Data

The F1/F2 score plot of the FTIR data in Fig. 7.14 again reveals an obvious rank (degree of coalification) trend as can be seen from the monotonously progressing sample numbers when projecting the various

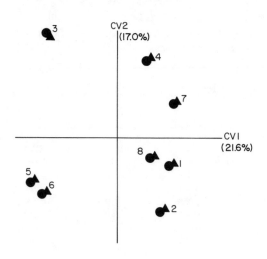

FIGURE 7.13. Superimposed score plots of LVMS and FIMS data in common CV1/CV2 spaces.

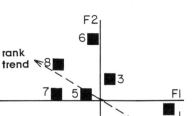

FIGURE 7.14. Factor score plots of FTIR data in the factor F1/F2 space.

scores on the presumptive rank trend axis. Considering the predominant chemical effect of increasing rank, i.e., a transition from oxygen-substituted compounds to aliphatic and aromatic hydrocarbons, and the expected changes in the corresponding FTIR spectra, the strong rank trend in Fig. 7.14 is not unexpected. At the same time, however, the FTIR score data are obviously rotated relative to the LVMS score data in Fig. 7.7. The rotation between the two data sets is also illustrated by the corresponding table of cross correlations (Table 7.4) in view of the relatively high correlations between several of the major principal components in both data sets.

7.3.6. Canonical Correlation of LVMS and FTIR Data

Table 7.5 illustrates the effect of canonical correlation analysis on the cross correlations between the first five principal components of the LVMS and FTIR data sets. Although a definite degree of diagonalization is achieved, the diagonal elements do not nearly approach the same high levels as in the LVMS/FIMS cross-correlation matrix. Consequently, the off-diagonal elements reveal a distinct degree of residual nonorthogonality. It is instructive to compare the principal component cross correlations in

TABLE 7.4. Correlation between LVMS and FTIR Principal Components

	LVMS				
FTIR	F1	F2	F3	F4	F5
F1	−0.686	0.735	−0.164	−0.105	−0.010
F2	0.524	0.171	−0.422	0.617	0.007
F3	0.575	0.422	−0.161	−0.570	−0.141
F4	−0.086	−0.758	0.150	0.245	−0.322
F5	−0.188	0.128	−0.431	0.467	−0.421

TABLE 7.5. Correlation between LVMS and FTIR CV Functions

FTIR	LVMS				
	CV1	CV2	CV3	CV4	CV5
CV1	0.942	−0.001	0.008	0.001	0.004
CV2	0.060	0.908	0.034	0.005	−0.004
CV3	−0.040	0.040	0.876	0.009	−0.018
CV4	−0.003	0.001	0.003	0.797	0.001
CV5	0.073	−0.099	0.043	0.006	0.767

Table 7.4 with those in Table 7.2. At first sight it would appear that Table 7.4 indicates a higher degree of correlation between the two data sets (LVMS vs. FTIR). For instance, Table 7.4 lists two correlation coefficients in the >0.700 range and four in the >0.600 range as compared to only half these numbers (one >0.700, two >0.600) for Table 7.2. Nonetheless, the rotated data in Tables 7.3 and 7.5, leave little doubt which two data sets are more closely correlated. This illustrates the need to search for optimally rotated combinations of principal components rather than to attempt a quantitative and/or qualitative interpretation of the original, nonrotated principal component axes, which merely represent directions of maximum total variance regardless of the strength of the correlations between the two data sets.

Upon performing canonical correlation analysis again a decision needs to be made regarding the number of CV functions that should be retained. The canonical correlation coefficients in Table 7.5 indicate the precence of only two or three dimensions in which strongly correlating scores can be observed. However, direct visual inspection of the CV scores in the first four dimensions (Fig. 7.15) suggests that a clear degree of correlated behavior is even present in the direction of CV4. Besides using the earlier-discussed statistical criteria (percent total variance explained, canonical correlation coefficients, Wilk's criteria, chi-square coefficients), one might want to consider an approach based on the underlying chemistry. In other words, one would like to find those dimensions of CV space that make good spectroscopic sense (i.e., that are chemically interpretable) as opposed to dimensions composed of more or less random collections of variables and thus likely to represent primarily chemical "noise." The variance diagrams corresponding to the CV1/CV2 and CV3/CV4 subspaces of the LVMS data set in Fig. 7.16 reveal the directions of the now familiar chemical components A–F (see Figs. 7.8 and 7.9), all of which appear to be represented in the CV1–CV4 subspace. For components A–D this is further illustrated by the numerically extracted component spectra in Fig. 7.17, which may be compared with the corresponding spectra in Fig. 7.9.

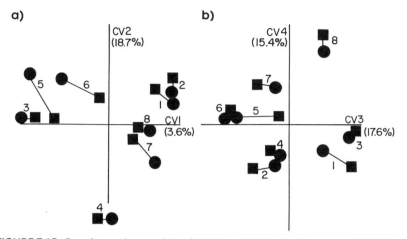

FIGURE 7.15. Superimposed score plots of LVMS and FTIR data in common CV1/CV2 (a) and CV3/CV4 (b) spaces, respectively.

Unfortunately, we were unable to produce the corresponding variance diagrams of the CV1–CV4 FTIR subspace due to limitations of our current version of the VARDIA program with regard to number of variables and data format. Thus, we decided to calculate FTIR component spectra in the same overall directions in CV space as used for the numerically extracted mass spectra in Fig. 7.17. The results, obtained directly with the CIRCOM program, are shown in Fig. 7.18 and thus may be compared directly with the corresponding mass spectra in Fig. 7.17. Good correspondence is

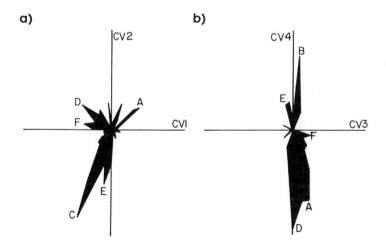

FIGURE 7.16. The variance diagrams corresponding to the CV1/CV2(a) and CV3/CV4(b) subspaces of the LVMS data.

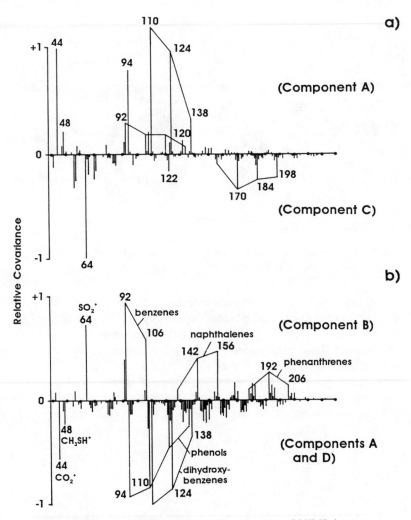

FIGURE 7.17. Numerically extracted CV components of LVMS data.

observed with regard to aromatic and heteroatomic signals. As expected, however, the FTIR spectra contain some components, e.g., representing mineral matter components, which appear to have no direct counterpart in the LVMS profiles, and *vice versa*.

Observation of mineral matter related FTIR signals in the CCA subspace poses an interesting philosophical problem: how can signals representing structural aspects of the sample which cannot be directly observed by one of the spectroscopic techniques end up within the most strongly correlating dimensions of CV space? Obviously, the correlations

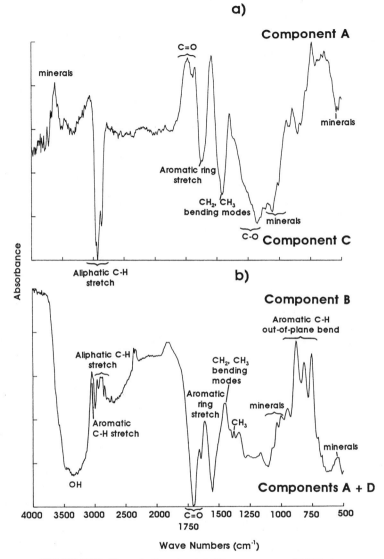

FIGURE 7.18. Numerically extracted CV components of FTIR data.

found must be due either to chance (quite conceivable when analyzing only eight coal samples) or to complex, hidden relationships between samples. For instance, a high organic sulfur content may occur in association with certain inorganic components due to a common cause, e.g., a particular depositional environment.

An annoying problem in Fig. 7.18 which complicates the assignment of the various peak signals is the uncertainty with regard to the position and shape of the baseline. Future adaptation of the VARDIA technique for use with FTIR data should be a key step toward solving baseline correction problems by enabling us to locate specific component vectors in FTIR data space. Experience with MS data indicates that the most characteristic components are often represented primarily by positive loadings, thereby obviating or reducing the need for baseline correction.

7.4. CONCLUSIONS

Principal component ("factor") analysis of each of the three data sets reveals a relatively high intrinsic dimensionality compared to the number of coal samples available. This seemingly unfavorable ratio does not appear to compromise the use of CCA in an exploratory data analysis mode, however. The first four CV functions calculated for the LVMS/FIMS data sets explain more than 70% of the total variance, while approximately 40% of the total variance in LVMS/FTIR CV space is explained by the first three CV functions.

The large fractions of total variance explained by the most significant three to five CV functions are interpreted as strong indications that the correlations found are unlikely to be due to noise components. Also, the natural tendency of CCA to reduce the influence of noise components by effectively applying a correlation filter will help to reduce the effect of non-correlated sources of variance. Further more, the fact that the chemical trends and components in CV loading space can be readily explained in terms of the known chemical composition and structure of the ANL coal sample set would seem to minimize the risk that the strong correlations between the data sets are based on chance observations.

Chemical interpretation of the trends and components in principal component space using the Variance Diagram (VARDIA) technique reveals the fact that the LVMS and FTIR data sets are dominated by a strong rank ("degree of coalification") trend. However, this trend is not obvious in the FIMS data set in spite of the fact that very similar chemical components are present in F1/F4 space.

Careful inspection of the CV1/CV2 and CV3/CV4 loading spaces uncovers the presence of familiar chemical components, e.g., representing coal rank as well as depositional environment. Finally, it should be noted that some of the correlations found involve chemical components which cannot be measured directly by one of the spectroscopic techniques. For instance, mineral matter signals, although without obvious counterpart in the EIMS or LVMS data, figure prominently in the numerically extracted FTIR spectra.

Obviously, far from being a "black box" technique for routine spectroscopic applications, CCA as described here should rather be received as a promising exploratory data analysis tool for multisource data. When used carefully, CCA should help analytical chemists to formulate the right questions, e.g., to be answered by the highly specialized expert spectroscopist. In our experience, CCA can provide a unique vehicle for cross-disciplinary interactions between spectroscopists who had never been able to communicate effectively before. As such CCA would seem to deserve its own place as a relatively transparent exploratory data analysis tool, next to conceptually more sophisticated but less transparent techniques such as PLS (Partial Least Squares), which tend to reach peak performance in the modeling and prediction stage of multisource data analysis.

REFERENCES

1. J. F. Hair, R. E. Anderson, and R. L. Tatham, *Multivariate Data Analysis*, 2nd ed., Macmillan, 1987.
2. P. E. Green and London, J. D. Carroll, *Mathematical Tools for Applied Multivariate Analysis*, Academic Press, New York, 1976.
3. R. P. Lattimer, J. B. Pausch, and H. L. C. Meuzelaar, *Macromolecules* **16**, 1896–1900 (1983).
4. W. Windig, H. L. C. Meuzelaar, B. A. Haws, W. F. Campbell, and K. H. Asay, *J. Anal. Appl. Pyrol.* **5**, 183–198 (1983).
5. W. Windig and H. L. C. Meuzelaar, Proc. 31st ASMS Conf., Boston, MA, pp. 893–894 (1983).
6. G. Halma, D. van Dam, J. Haverkamp, W. Windig, and H. L. C. Meuzelaar, *J. Anal. Appl. Pyrol.* **7**, 167–183 (1984).
7. W. Windig, H. L. C. Meuzelaar, F. Shafizadeh, and R. G. Kelsey, *J. Anal. Appl. Pyrol.* **6**, 233–250 (1984).
8. A. M. Harper, H. L. C. Meuzelaar, and P. H. Given, *Fuel* **63**, 793–802 (1984).
9. H. L. C. Meuzelaar, W. Windig, and G. S. Metcalf. *Am. Chem. Soc., Div. Fuel Chem., Prepr.* **30**(1), 200–208 (1985).
10. J. L. Savoca, R. P. Lattimer, J. M. Richards, W. Windig, and H. L. C. Meuzelaar, *J. Anal. Appl. Pyrol.* **9**, 19–28 (1985).
11. W. Windig, M. Nip, S. Beckman, Karl Schorno, and H. L. C. Meuzelaar, Proc. 33rd ASMS Conf. on Mass Spectrom. All. Topics, San Diego, CA, May 26–31, 257–258 (1985).
12. J. L. Savoca, R. P. Lattimer, H. L. C. Meuzelaar, and J. M. Richards, Proc. 33rd ASMS Conf. on Mass Spectrom. All. Topics, CA, May 26–31, San Diego, pp. 806–807 (1985).
13. H. L. C. Meuzelaar, B. L. Hoesterey, W. Windig, and G. R. Hill, *Fuel Processing Technology* **15**, 59–70 (1986).
14. K. Taghizadeh, R. Hardy, R. Keogh, J. Goodman, B. Davis, and H. L. C. Meuzelaar, *ACS preprint*, 194th National Meeting, New Orleans, August 31–September 4, **32**(3), 332–341 (1987).
15. W. Windig and H. L. C. Meuzelaar, in: *Computer-Enhanced Analytical Spectroscopy*, Volume 1 (H. L. C. Meuzelaar and T. L. Isenhour, eds.), pp. 67–102, Plenum Press, New York (1987).

16. T. Chakravarty, H. L. C. Meuzelaar, P. R. Jones, and R. Kahn, *Am. Chem. Soc., Div. Fuel Chem., Prepr.* **33**(2), 235–241 (1988).
17. G. J. C. Yeh, B. Ward, D. R. Quigley, D. L. Crawford, and H. L. C. Meuzelaar, *ACS Preprint*, 196th National Meeting (revised), Los Angeles, CA, September 25–30, **33**(4), 612–622 (1988).
18. B. L. Hoesterey, W. Windig, H. L. C. Meuzelaar, E. M. Eyring, D. M. Grant, and R. J. Pugmire, in: *Pyrolysis Oils from Biomass: Producing, Analyzing, and Upgrading*, ACS Symposium Series **376** (J. Soltes and Thomas A. Milne, eds.), pp. 189–202 (1988).
19. K. Taghizadeh, H. L. C. Meuzelaar, and H. J. Hatcher, *37th ASMS Conf. on Mass Spectrom. All. Topics*, Miami Beach, Florida, May 21–26, pp. 302–303 (1989).
20. B. L. Hoesterey, H. L. C. Meuzelaar, and R. J. Pugmire, *Energy and Fuels* **3**, 730–734 (1989).
21. K. Taghizadeh, B. H. Davis, W. Windig, and H. L. C. Meuzelaar, in: *Fossil Fuel Analysis by Mass Spectrometry* (T. Ash and K. Wood, eds.), ASTM, pp. 144–158 (1989).
22. K. Taghizadeh, R. R. Hardy, B. H. Davis, and H. L. C. Meuzelaar, *Fuel Processing Technology* (submitted).
23. M. R. Khan, T. Chakravarty, and H. L. C. Meuzelaar, *Ind. Eng. Res.* **29** (11), 2173–2180 (1990).
24. H. Hotelling, *J. Educ. Psychol.* **26**, 139 (1935).
25. H. Hotelling, *Biometrica* **28**, 321 (1936).
26. P. Horst, *Psychometrica* **26**(2), 129 (1961).
27. J. D. Carroll, Proceedings of the 76th Annual Convention of the APA, pp. 227–228 (1968).
28. M. S. Bartlett, *J. R. Stat. Soc.* **9**, 176 (1947).
29. D. Stewart and W. Love, *Psychol. Bull.* **70**(3), 160 (1968).
30. M. I. Alpert and R. A. Peterson, *J. Mark. Res.* **9**, 187 (1972).
31. K. S. Vorres, *Users Handbook for the Argonne Premium Coal Sample Program*, Argonne National Laboratory, Argonne, Illinois (1989).
32. E. Jakob, W. Windig, and H. L. C. Meuzelaar, *Energy and Fuels* **1**, 161–167 (1987).
33. E. Jakob, B. L. Hoesterey, W. Windig, G. R. Hill, and H. L. C. Meuzelaar, *Fuels* **67**, 73–79 (1988).
34. H. R. Schulten, *J. Anal. Appl. Pyrol.* **12**, 149–186 (1987).
35. H. R. Schulten, N. Simmleit, and R. Muller, *Anal. Chem.* **59**, 2903–2908 (1987).
36. H. R. Schulten and H. D. Beckey, *Org. Mass Spectrom.* **6**, 885–895 (1972).
37. W. Windig and H. L. C. Meuzelaar, Proc. 34th ASMS Conf. on Mass Spectrom. All Topics, Cincinnati Ohio, p. 64 (1986).
38. J. H. Wilkinson and C. Reinsch, "Linear Algebra," in: *Handbook for Automatic Computation*, Vol. III (F. L. Bauer et al., eds.), Springer-Verlag, New York (1971).
39. W. Windig, P. G. Kistmaker, and J. Haverkamp, *J. Anal. Appl. Pyrol.* **31**, 199–212 (1981).
40. H. L. C. Meuzelaar, H. Huai, R. Lo, and J. P. Dworzanski, *Fuel Processing Technology* (1990).

Developing Knowledge-Based Systems for Interpreting Infrared Spectra

8

Sterling A. Tomellini, Barry J. Wythoff, and Steven P. Levine

8.1. INTRODUCTION

It is often difficult to determine from the primary literature the philosophical basis, similarities, and differences between related knowledge-based systems. Work in our laboratories has concentrated on developing a number of knowledge-based systems for interpreting infrared spectra, for both organic functional group analysis and analysis of mixtures. These systems are at first glance separate and distinct entities. The development process for the systems described, even though they may have very different applications, often incorporates modifications which address deficiencies observed in the earlier systems. The goal of this chapter is to explain the recent improvements made to PAIRS (Program for the Analysis of Infrared Spectra) and the development of several systems which have their basis in PAIRS.

Sterling A. Tomellini and Barry J. Wythoff • Department of Chemistry, University of New Hampshire, Durham, New Hampshire 03824-3598. Steven P. Levine • Department of Environmental and Industrial Health, School of Public Health, University of Michigan, Ann Arbor, Michigan 48109-2029.

Computer-Enhanced Analytical Spectroscopy, Volume 3, edited by Peter C. Jurs. Plenum Press, New York, 1992.

8.2. ORGANIC FUNCTIONAL GROUP ANALYSIS—PAIRS

Much of the work in our laboratory has concentrated on developing and improving PAIRS,[1-6] which is a rule-based expert system that has undergone continuous development for approximately ten years. The goal in developing PAIRS was to assist chemists in determining which functionalities and subfunctionalities are likely to be present in an unknown compound, based on its IR spectrum. The program attempts to emulate the problem-solving approach taken by a trained IR spectroscopist and reports the results of the interpretation to the chemist. At the present time, interpretation rules have been written for over 195 organic functionalities and subfunctionalities.

The PAIRS interpreter requires three sources of information to perform a spectral interpretation. It requires the spectral information, which is represented by a list of each peak's position $(4000-400 \text{ cm}^{-1})$, intensity $(1-10)$, and width (sharp, medium, or broad). The user may also provide PAIRS with additional information such as an empirical formula, if known, for the compound and the sampling conditions. During an interpretation, the system accesses a file which represents the knowledge and logic that a spectroscopist might use to interpret an IR spectrum.

The knowledge and logic are coded into a large set of IF-THEN-ELSE type rules, which are also commonly termed "production rules." These rules form the knowledge base of the system. The rules were developed by translating decision trees which were created for each functionality and subfunctionality based on information contained in IR correlation charts and other publications. One philosophical question which needed to be addressed when developing these decision trees was whether the decisions being made should be biased toward avoiding false positive results (i.e., indicating a functionality is likely to be present when it actually is not) or avoiding false negative results. It was rationalized that since the user has to make the final decision as to the likelihood of presence for a given functionality, then the system should be biased toward avoiding false negative results. It is important that the user (or anyone evaluating the system's performance) understand that the rules, and therefore the interpretation results, are biased in this way.

Another question which needed to be addressed early during the development of the system was what information should constitute the results of a PAIRS interpretation. It was decided at that time that the interpretation results could be adequately presented by a simple list containing the numerical likelihood of presence for each functionality type. Thus, the results of an interpretation session for the earliest versions of the system were limited to a list of expectation values, one for each functionality or subfunctionality in the rule base. The expectation values which were presented ranged from a low of 0.01 (indicating a very low likelihood of presence) to a high of 0.99 (indicating that spectral data suggest a very high likelihood for the presence of the functionality in question). It should be noted that the interpreter was not allowed to report an expectation value of 0.00 or 1.00, indicating that a given functionality was definitely present or absent, respectively. This was done to emphasize to the users that they, not PAIRS, must make the final decision concerning the presence of a given functional group.

The interpretation of a spectrum of a simple compound, ethylbenzene, will be used to demonstrate the capabilities and limitations of an early version of PAIRS. A spectrum of ethylbenzene was acquired as a thin film between salt plates and is presented in Fig. 8.1. The spectrum was peak picked. The intensities of each peak were normalized on a scale from 1 to 10 by dividing by the absorbance of the band at $697\,cm^{-1}$, which is the most intense peak in the spectrum. The peak position, intensity, and width data which were entered into PAIRS are given in Table 8.1. These data were interpreted by PAIRS without having access to any empirical formula information. The interpretation results, as provided by an early version of the system, are presented in Table 8.2.

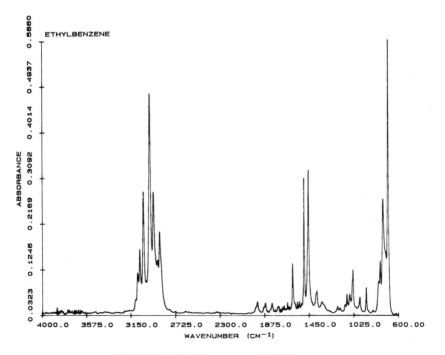

FIGURE 8.1. The IR spectrum of ethylbenzene.

TABLE 8.1. Peak Position, Normalized Intensity, and Width Data
for the Spectrum of Ethylbenzene[a]

	Peak position	Normalized intensity		Peak position	Normalized intensity
(1)	697	10	(15)	1558	1
(2)	745	5	(16)	1604	2
(3)	771	3	(17)	1651	1
(4)	903	2	(18)	1802	1
(5)	965	1	(19)	1865	1
(6)	1031	2	(20)	1941	1
(7)	1063	1	(21)	2873	4
(8)	1089	1	(22)	2894	3
(9)	1109	1	(23)	2931	6
(10)	1328	1	(24)	2965	9
(11)	1376	2	(25	3027	5
(12)	1454	6	(26)	3063	3
(13)	1495	5	(27)	3084	2
(14)	1540	1			

[a] All of the bands have "average" width.

When presented with the results of an interpretation such as those given in Table 8.2, the developers of PAIRS needed to address a number of issues. Two of the most important questions were: (1) How could the user determine the reasons for the expectation value reported for a given functionality, and (2) How could system performance be improved? Initially, these questions were viewed as having separate and distinct answers. It seemed clear that to improve system performance, the key was to improve, refine, or expand the interpretation rules. It was reasonable to believe that "better" interpretation rules would provide better interpretation results. Much of the effort in improving PAIRS at that point was thus directed toward this end by interpreting test spectra, analyzing the results, and subsequently modifying the rules. It became obvious, however, that the quality of an interpretation did not rest solely with the quality of the interpretation rules.

Several manual revisions of the rule base were undertaken to improve the interpretation results provided by the system. Often, what appeared to be a "bad" interpretation result was actually reasonable, based on the spectral data provided to the system. Based on this observation, it became apparent that the main limitation of the system at this point was not the rule base, but instead the quantity and quality of information which the system was providing to the user. A numerical list of expectation values, as given in Table 8.2, simply does not provide the user with all the infor-

TABLE 8.2. The Interpretation Results as Provided
by an Early Version of PAIRS for the Spectral Data
Given in Table 8.1 [a]

	Expectation Results with Original Peak List	
Functionality		Expectation value
(1) AROMATIC		0.95
(2) THIOPHENE		0.90
(3) METHYL		0.65
(4) HETEROAROMATIC		0.50
(5) PYRIDIL		0.50
(6) METHYLENE		0.40
(7) AMINE		0.40
(8) AMINE-TERTIARY		0.40
(9) AROMATIC-1.3-SUBSTITUTED		0.29
(10) AROMATIC-MONOSUBSTITUTED		0.29
(11) AROMATIC-1.2-SUBSTITUTED		0.29

[a] All remaining functionalities and subfunctionalities have expectation values below 0.29.

mation required from the interpretation process. It became clear that to improve PAIRS further, the user must be informed of the decisions which the system was making to arrive at its conclusions. For example, the user might be interested in knowing what information was being used to arrive at an expectation value of 0.95 for the functionality "AROMATIC" in the above interpretation of the spectral data for ethylbenzene. PAIRS was modified to provide such information by allowing the user to trace through the decision making process in an automated fashion.[5] An example of such a decision trace is given in Appendix 8.1 for the "AROMATIC" functionality.

While there is little doubt that the value of a PAIRS interpretation was increased by allowing the user to trace the decision-making process, it is also clear that the system was still providing significantly less information than would normally be provided by an infrared spectroscopist. In the ethylbenzene example, the user could now easily determine what spectral data were responsible for the high expectation value for "THIOPHENE" but was not able to determine the system's rationale for querying about those data. To overcome this limitation, PAIRS was again modified, this time with the goal being to provide the user with the rationale behind the interpretation process.[6] Approximately 2000 lines of explanatory comments were added to the PAIRS rule base to facilitate this change. An example of this type of information now available to the user during the interpretation of the data given in Table 8.1 is presented in Appendix 8.2 for the functionality "AROMATIC-MONOSUBSTITUTED."

The user, when provided with this additional information, is capable of reviewing the spectral data and is free to agree or disagree with the desisions made by the interpreter and/or the reasons for making those decisions. The informed user can, for example, note that the final query for "AROMATIC-MONOSUBSTITUTED" asks if there is an absorption band with intensity greater than 6 between 721 cm^{-1} and 820 cm^{-1}. The answer to this query is "no" based on the spectral data, but the user may want to question this answer based on the presence of the band at 745 cm^{-1} having an intensity of 5. To facilitate such interaction between the user and the system, PAIRS was further modified to allow for an interactive interpretation process where the user is allowed to override the decisions being made by the interpreter. Appendix 8.3 demonstrates the use of the interactive interpretation option for the "AROMATIC-MONOSUBSTITUTED" functionality. If the user overrides the "no" decision made by the interpreter based on the intensity of the band at 745 cm^{-1}, the resulting expectation value for "AROMATIC-MONOSUBSTITUTED" is 0.86.

The reader should note that the role of the user has undergone a fundamental change during the development of PAIRS. Originally, PAIRS was designed to provide the user with an answer to the question, "Is this functionality likely to be present in this sample based on the spectral data?" The latest version of the system can, however, supply the user with the answer to the question, "Why is this functionality likely to be present (or absent) based on the spectral data?" Other versions of PAIRS have also been developed to allow the user to participate interactively in the interpretation process, including one version which is graphically oriented.[7]

8.3. USING EXPERT SYSTEMS FOR INTERPRETING SPECTRA OF MIXTURES

8.3.1. PAWMI

Based on the early success of PAIRS for functional group analysis and the availability of the PAIRS system, Levine's group at Michigan became interested in using a rule-based expert system for determining the likelihood of the presence of specific compounds in samples found at hazardous waste sites.[8] A system based on PAIRS and named PAWMI was created to interpret such spectra.[9–11]

In a mixture, it is expected that bands which are present in the spectrum of a pure compound will be present in the spectrum of the mixture if the compound is a component of that mixture. It is known, however, that the positions of the absorption bands for a compound are often shifted from those in the spectrum of the pure compound for a number of reasons, including matrix effects and peak picking inaccuracies due to the presence of other absorption bands in the same region. It is also known that absolute band intensities are of little diagnostic value when interpreting spectra of mixtures. Two reasons for this are: (1) molar absorptivities are often matrix-dependent, and (2) the concentration of each component will vary from one sample to the next. Based on these observations, the challenge was to develop interpretation rules for specific compounds that would replace the PAIRS functional group rules.

The first step in creating the interpretation rules for each compound in the rule base was to determine the ten largest bands in the spectrum of the compound. If the compound's spectrum had at least ten bands without using those in the carbon–hydrogen stretching region, then bands in this region were eliminated from consideration. This action was taken since most organic compounds contain bands in this region of the spectrum, thereby limiting their diagnostic value. Once the positions of these so-called "rule peaks" were determined, interpretation rules could be created. The expectation values in PAWMI were allowed to range from 0.0 to 1.0. Each of the rule peaks was weighted equally, thus the fraction of the total expectation value allocated for each rule peak was simply 1.0 divided by the number of rule peaks for the compound. If a compound had ten rule peaks, then each was worth a maximum of 0.1, or 10% of the total possible expectation value of 1.0. Three concentric window regions were then centered about each peak position to account for peak shifting, which is expected in mixtures. The windows chosen were $\pm 3 \, \text{cm}^{-1}$, $\pm 5 \, \text{cm}^{-1}$, and $\pm 10 \, \text{cm}^{-1}$ about the peak position of each of the rule peaks for the compound. The value of each rule peak was further apportioned between these subregions. If a band in the spectrum was within the $\pm 10 \, \text{cm}^{-1}$ region about the rule peak, then 20% of the value available for that rule peak was assigned. If the peak in the mixture was within the $\pm 5 \, \text{cm}^{-1}$ window, then 50% was assigned, and if the peak in the mixture fell within the $\pm 3 \, \text{cm}^{-1}$ region about the rule peak, then 100% of the value available for that rule peak was assigned. It should be emphasized that the regions and numerical values were chosen empirically. An example of the decision tree generated for a single rule peak occurring at $1700 \, \text{cm}^{-1}$ is presented in Fig. 8.2.

Interpretation rules like those described above were written for 62 of the compounds most frequently encountered at hazardous waste sites. These rules are capable of providing the user with a numerical indication of the spectral similarity between the absorption bands found in the spectrum of the pure compound and the bands found in the spectrum of the mixture. Thus, they provide the user with an answer to the question,

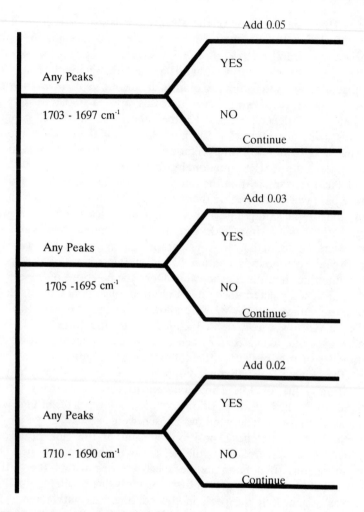

FIGURE 8.2. An example of a PAWMI decision tree for one absorption band in the pure compound. The absorption band occurs at $1700\ cm^{-1}$. The total expectation value assigned for this band is assumed to be 0.10.

"Is it likely that a given compound is present in the mixture based on its spectrum as a pure compound?"

One problem which occurs when using the aforementioned rules for interpreting spectra is that chemically similar compounds will have very similar infrared spectra. It is expected, therefore, that if a given compound is present in a mixture then one would expect the spectral data to suggest that any similar compounds are also somewhat likely to be present in the mixture.

PAWMI employed a post-interpretation algorithm, known as "PAIRSPLUS," to enhance the system's ability to distinguish between spectrally similar compounds. PAIRSPLUS was designed to allow the user to employ the results of the compound-specific interpretation to answer the question, "Which of the compounds having high expectation values are most likely to be present in the mixture after taking into account spectral similarity?"

The approach taken was to interpret the spectrum of each of the pure compounds using the interpretation rules of the other compounds contained in the rule base. Clearly, the closer the spectral similarity between two compounds, the higher the expectation value which would result from interpreting one compound's spectrum with the other compound's rules. The results of these interpretations were then placed in a file which contains the 62×62 matrix of expectation values for later use by the PAIRSPLUS routines.

PAIRSPLUS assumes that the lower expectation values for compounds are primarily due to their spectral similarity with those compounds having higher expectation values. It assumes, therefore, that the compounds having high expectation values. are more likely to be present in the sample than those having lower expectation values. It is important to note that the inherent assumption in this approach is that *both* compounds are not likely to be present. PAIRSPLUS modifies the expectation value scores in a sequential manner starting with the compound with highest expectation value, to account for the spectral similarity of the compounds using the previously described matrix.

PAWMI was found to have two significant limitations which needed to be addressed in future systems. The first was that the rules used for PAWMI were manually generated in the CONCISE language used by PAIRS. Generating interpretation rules in this manner is both tedious and time-consuming. The second problem was that only peak position information was used to create the interpretation rules used in PAWMI. It was postulated, as during the development of PAIRS, that the key to improved, compound-specific interpretations was the development of better interpretation rules. One way to solve both problems was to develop an automated rule generation program designed to incorporate more spectral information into the rules created. The use of an automted rule generation program, which created PAIRS interpretation rules for functional group analysis, had been demonstrated previously.[4]

8.3.2. intlRpret

Prior to developing an automated generation program for compound specific rules, it was necessary to determine what information could be utilized during rule development while still maintaining a simple rule structure. It was decided that one solution to the problem was to devise a scheme for estimating the relative diagnostic value of reference bands, and the relative likelihood of observing them in a mixture. In creating the PAWMI interpretation rules, each rule peak in the spectrum of the pure compound was allocated the same maximum expectation value based simply on the number of rule peaks used for that compound. This simple method of peak weighting was used, since it allows for more rapid rule generation.

A more sophisticated approach to allocating the maximum weights for each rule peak was incorporated into the automated rule generation program. The approach takes into account three factors which are based on the spectral data, but continued to use the same window regions (± 3 cm^{-1}, ± 5 cm^{-1}, ± 10 cm^{-1}) and relative weightings for each window (100%, 50%, and 20%, respectively). The first factor is based on the intensity of the rule peak relative to other rule peaks for the compound. It is expected that the more intense absorptions of the pure compound are more likely to be observed in a mixture containing that compound and, therefore, these peaks should be weighted more heavily. The second factor considers the intensity of the peak in the spectrum of the pure compound relative to other peaks within the same wavenumber region from the other compounds in the data base. Those peaks in a compound which occur in a region containing few peaks due to other compounds are more indicative of the compound in question, and therefore should be weighted more heavily. The third factor used to determine the maximum expectation value to allocate for a given rule peak is the number of rule peaks for the compound within a given wavenumber window relative to the total number of rule peaks for the other compounds in the data base within that wavenumber window. These latter two factors are related to the peak "occurrence" factor, which was incorporated into a PAIRS automated rule generation program that had been created previously.[4]

The resulting system, which included the enhanced interpretation rules created by the automated rule generator, was named "intIRpret."[12,13] Comparison of the interpretation results for the same test data set for PAWMI and intIRpret proved that the additional logic used in developing the intIRpret rules provided for better interpretation results.

8.3.3. IRBASE/MIXIR

At this point the strengths and weaknesses of the PAIRS, PAWMI, and intIRpret systems were reevaluated to determine how to further advance the use of knowledge-based systems for interpreting IR spectra of mixtures. The evaluations revealed that one of the primary limitations of all of these systems is the structure of the interpretation rules. The knowledge and logic used for a spectral interpretation are fixed and contained in a set of hard coded rules. It would clearly be difficult, if not impossible, to generate rules of this type which are capable of handling all situations likely to be encountered when dealing with unknown mixtures. For this reason, it would be desirable to have a knowledge base in which both the knowledge and logic can be varied to suit the interpretation problem at hand. Since both the spectral knowledge and interpretation logic are hard coded in PAIRS-type rules, they cannot be modified easily during the interpretation process. Another problem with using PAIRS-type interpretation rules is that it is difficult to create rules with a complex logical design using an automated rule generator like those previously developed for PAIRS and intIRpret.

The solution to the problem caused by the form of the PAIRS rules was to separate the spectral knowledge from the logical approach to be used during the interpretation process. Toward this end, we decided to develop two interdependent systems, IRBASE and MIXIR, for interpreting IR spectra of mixtures.[14,15]

The first system, IRBASE, creates the compound-specific spectral descriptions which are used by MIXIR to interpret the spectra of unknown mixtures. The goal in developing IRBASE was to generate the compound-specific spectral descriptions in an automated manner, using spectral and chemical information about the compound. To accomplish this goal the amount of information used to create the spectral descriptions for each compound was increased. For example, we desired to take a more sophisticated approach than used previously for establishing the position of window regions to be used about each absorption band in the spectrum of the pure compound. Previously, windows of $\pm 10 \, \text{cm}^{-1}$, $\pm 5 \, \text{cm}^{-1}$, and $\pm 3 \, \text{cm}^{-1}$ were used without regard to the functional group that gave rise to the absorption band. IRBASE uses information provided by the user about the functional groups that comprise the compound in order to determine which observed absorptions are due to which functional groups. Reasonable limits for band shifting for the most commonly encountered functional groups have been incorporated into the program and are used to assign windows for each band.

IRBASE also utilizes peak intensity and width information to create the spectral descriptions for each compound. The absolute intensity of a band in the pure compound is, however, still of limited use when creating the compound's spectral description. In contrast, relative peak intensities are expected to have diagnostic value. As with the intIRpret autogenerator, IRBASE uses the relative peak intensities for the compound, since it is expected that the more intense absorptions of a compound are more likely to be observed in the spectrum of the mixture while the smaller absorption bands for the compound would not be expected to be observed.

IRBASE uses an empirical algorithm to generate two separate spectral descriptions for each compound to be included in the knowledge base. One spectral description is used to determine if it is likely that the compound is present as a major component in the sample. The second spectral description determines if the compound is a minor component of the mixture. Each spectral description contains peak position and intensity information for that compound. IRBASE also indicates which absorptions are essential and must be observed in the spectrum of the mixture for the compound to be present. These descriptions along with a total peak file listing for all of the compounds in the knowledge base are used by MIXIR. IRBASE has been used to create spectral descriptions for 50 condensed-phase compounds using standard library spectra. Many of the compounds included in the knowledge base were chosen due to their ability to hydrogen-bond. The compound descriptions have been used to test the capabilities of the MIXIR system.

MIXIR is the spectral interpretation program. It was designed to involve the user in the interpretation process. MIXIR accesses the compound-specific spectral descriptions previously compiled by IRBASE. These spectral descriptions provide the knowledge necessary to interpret the spectrum of the mixture. Separating the spectral descriptions from the logic used by the system provides the significant advantage of being able to manipulate all the information as needed. MIXIR, because of this structure, can approach a spectral interpretation in a dynamic manner. It can, for example, update the importance given to finding or missing a spectral feature of a compound based on other information which it has "learned" during the interpretation process.

The logic used during an interpretation session is not contained in hard coded IF-THEN-ELSE type rules like those which were used in PAIRS and PAWMI. Instead, MIXIR contains a number of logical paradigms that can be chosen to perform the interpretation. The user chooses which logical options to employ for the interpretation and may participate directly in the interpretation process if so desired. Thus, the MIXIR system plays the role of a "smart assistant" to the user during an interpretation session, rather than that of a replacement for the user.

The user is asked to choose among nine interpretation options and paradigms prior to performing an interpretation. Any or all of these options may be used to interpret a given spectrum. The options are: user override, relative peak intensity checking, essential peak checking, use of an extended scoring system, automatic cycling, a trace of the interpretation, peak justification, reduced peak set checking, and peak swamping check.

The user override option clearly shows the influence of the latest user-interactive version of PAIRS on the development of MIXIR. This option allows the user to participate in the interpretation process at the most basic level, overriding decisions and actions made by the MIXIR interpreter if so desired.

The peak justification option is used when the user wants to make sure that all of the most intense peaks in the spectrum of the unknown mixture can be attributed to one of the compounds which is interpreted to have a reasonably high likelihood of being present. Through a separate, non-optional procedure, if none of the compounds can account for the presence of a band in the spectrum of the mixture, then the user is informed that it is likely that a compound not included in the MIXIR set of compounds is probably present in the mixture.

The automatic cycling option is probably the best demonstration that separating the spectral descriptions from the logic in this system provides the significant advantage of flexibility in the way the information contained in the descriptions can be used. MIXIR, because of this structure, is capable of approaching a spectral interpretation dynamically, updating the importance of finding or missing spectral features in the spectrum of the mixture. This option provides for an automated reinterpretation of the spectrum, adjusting the values assigned for finding or missing a spectral feature of the pure compound based on the results from the previous pass of the interpreter. Thus, the system essentially "learns" as it interprets the spectrum of the mixture with the goal being to arrive at the best interpretation results possible, based on the specifics of the spectrum being interpreted.

The user can also choose among three post-interpretation options to obtain additional information. These three options are: an additional peak assignment option, an either/or procedure, and an interpretation cycle option that is used in conjunction with the automatic cycling option described above. The either/or procedure also demonstrates the flexibility provided by the separation of spectral descriptions from logic. This option is used to distinguish between two compounds which have similar spectra. MIXIR, using the spectral descriptions for each compound that have been compiled by IRBASE, derives a spectral description for each compound which disregards the spectral features that the two compounds have in common. The system then reinterprets the spectrum of the mixture to determine which of the two compounds is more likely to be present based on spectral data from the mixture.

When interpreting an IR spectrum of a mixture, MIXIR must determine the significance of finding features in the mixture spectrum that appear to correspond to "reference" features from the spectrum of the pure compound. Likewise, MIXIR must also determine the significance of any spectral data which are expected for the compound and are found to be missing from the mixture's spectrum. The presence and absence of spectral information for a given compound is used to arrive at a numerical score, which indicates to the user the likelihood of a particular compound being a component of the mixture. MIXIR scores range from a high of $+0.999$ (which indicates there is a very high likelihood that the compound is a component of the mixture) to a low of -0.999 (which indicates virtually no information suggests the compound is present in the mixture). These scores are calculated using a complex algorithm based on the observation that if a peak is observed in the spectrum of the mixture that corresponds to an absorption of the pure compound, then a positive adjustment must be made to the score for that compound. The magnitude of the adjustment must correspond to the belief that the band in the spectrum of the mixture is actually due to the presence of the compound in question, rather than another compound which might be a competing explanation. Similarly, if a pure compound has an absorption in a given wavenumber region and a corresponding absorption is not observed in the spectrum of the mixture, then a negative adjustment must be made to the score of that compound. In this case, the magnitude of the negative adjustment must take into account the likelihood that the band is actually present, but is masked by a stronger absorption due to another component or that the band is too weak to be observed due to the concentration of the compound in the mixture. Many of these same concepts were used in the intIRpret automated rule generator to arrive at the expectation values assigned for a given absorption region. The key to the MIXIR interpretation system is that the weights are not fixed as in intIRpret, but can instead be dynamically modified as a spectrum is interpreted and additional information becomes available during the interpretation.

TABLE 8.3. Partial MIXIR Interpretation Results for a Mixture of 4:1 w/w Chlorobenzene/Ethylbenzene

Compound	Major	PM	PS	Minor	PM	PS
CHLOROBENZENE	0.999	12	12	−0.999	0	8
3-METHYLPENTANE	0.199	3	8	0.309	3	5
ETHYLBENZENE	−0.291	5	15	−0.714	1	5
DICHLOROMETHANE	−0.316	2	7	−0.999	0	2
m-DICHLOROBENZENE	−0.370	6	15	−0.496	3	10
TOLUENE	−0.558	3	12	−0.999	0	6
⋮						
CHLOROFORM	−0.999	0	4	−0.999	0	2

The example in Table 8.3 demonstrates the use of MIXIR for interpreting the spectrum of 4:1 w/w mixture of chlorobenzene and ethylbenzene, when employing the autocycle, essential peak checking, extended scoring system, and reduced peak checking options simultaneously. This combination of interpretation options was found to be "optimum" based on previous work.

The low score of ethylbenzene can be attributed to the high degree of spectral similarity between that compound and chlorobenzene, which is present in the test mixture at four times the concentration. The logical approaches employed by MIXIR have been biased toward producing false positive rather than false negative results. For this reason, interpretation options have been included that assist the user in not overlooking the presence of compounds which are possibly present in the mixture. If, for example, the peak justification option is employed during an interpretation of this mixture, the following messages would be presented to the user after the scores for each compound were reported: "The unknown peak at: 1453 cm^{-1} can't be attributed to any of the compounds above 0.20." The message is meant to alert the user that a compound having a lower expectation value is probably present in the mixture. The system also indicates to the user which of the compounds included in the MIXIR data base are possibly responsible for the observed absorption by reporting, "The following two compounds are the most likely explanations: ETHYLBENZENE, N-HEXANE." It should be noted that the performance of systems such as MIXIR is difficult to characterize, since the type of result described here would not appear in a false negative/false positive summary report. Still, providing such information to the user is in fact a very useful way of avoiding false negative results during actual use.

8.4. CONTINUING RESEARCH—USING MIXIR FOR INTERPRETING IR SPECTRA OF VAPOR-PHASE MIXTURES

Infrared analysis of organic vapors has many potential applications, including on-site measurement of toxic compounds at hazardous waste sites and in the work place. Levine found that, due to the spectral and chemical noise observed in the spectra of low-concentration vapor-phase species, the intIRpret system was of limited value for interpreting such spectral data. To overcome the limitations of intIRpret, Levine and co-workers successfully employed least-squares fitting (LSF) and iterative least-squares fitting (ILSF) techniques to the analysis of vapor-phase mixtures with the goal being to determine the likelihood of presence for specific compounds.[16,17]

The approach taken when using LSF techniques is to fit portions of the spectrum from the pure components to the spectrum of the mixture. This approach is fundamentally different from that used by systems such as PAIRS, PAWMI, intIRpret, and MIXIR, all of which are peak-based methods. It should also be noted that all of these four systems were initially developed for and tested on condensed-phase species.

One of the fundamental limitations of peak-based interpretation methods is obviously that the peaks in the spectrum of the mixture must be "observed" or the system does not have data to interpret. Analysis of the MIXIR interpretation results for spectra of vapor-phase species at concentrations ranging from 1 do 50 ppm suggested that the main limitation for interpreting these spectral data was the limited number of peaks which were picked in the spectra of the mixtures. The inability to determine peak locations in the spectrum is one of the fundamental limitations of using a peak-based method. For this reason we have concentrated on improving the peak list provided to MIXIR.

It was determined from early testing of the MIXIR system that the quality of the interpretation results was influenced greatly by the inclusion of small bands in the peak list of the spectrum of the mixture. To facilitate including such bands, a program was written which enables the user to specify the baseline in specific regions of the spectrum. This then allows a lower peak-picking threshold to be utilized and thus provides access to the smaller spectral features. It was determined that reducing the peak-picking threshold from 2.5% of the maximum absorption, which was used when peak-picking condensed-phase data, to 1.0% of the maximum absorption clearly produced better interpretation results. Using the baseline correction routines and allowing the user to set the peak-picking threshold interactively provided even better interpretation results. These results clearly illustrate the critical importance of the quality of the information provided to the MIXIR system. Additional research has been undertaken to explore the use of a neural network for peak verification and/or peak determination. Initial results indicate that the neural network peak verification/ determination program may improve the quality of the peak list used by MIXIR.[18]

Work is also continuing on modifying MIXIR for interpreting vapor-phase spectra. Vapor-phase features in mixtures should appear at the same positions as in the pure compound spectra. Therefore, basing the band-position windows on the functional group which give rise to the absorption, as was done by IRBASE for compounds in the condensed phase, is not reasonable. Instead, band-position windows for vapor-phase species are assigned to accommodate for any uncertainty in determining the location of the band position. The position windows were chosen empirically to be 3, 5, 10, and 25 cm^{-1} for bands which were deemed to be very sharp, sharp, average, and broad, respectively. Another modification which has been investigated is including real-valued intensities for the spectral descriptions of vapor-phase compounds instead of the integral intensity values used for condensed-phase spectral descriptions. The rationale behind this change is that changes in molar absorptivities for the isolated molecules are not expected to vary for vapor-phase species in a mixture. Also, vapor-phase spectra can easily be obtained with a fixed path length, thereby allowing absolute intensities to be of use during the interpretation.

We are also investigating the development of a version of MIXIR that does not use spectral descriptions compiled by IRBASE. IRBASE, as originally designed and developed for condensed-phase data, essentially served to reduce the spectral features of the pure compound to prevent the dilution of the significance of important spectral features during an interpretation. One of the reasons for preprocessing these data was to improve execution speed. It is now felt that the optimum spectral description for each compound should be created dynamically during the interpretation process. The entire set of spectral features for each compound must be exploited using whatever information is known about the sample in question. Delaying the decision-making process as long as possible should enhance the ability of MIXIR to interpret adaptively an unknown sample. It is envisioned that these changes in system structure will enhance the ability of MIXIR to interpret IR spectra of vapor-phase mixtures.

8.5. CONCLUSION

Work in our laboratories has resulted in the development of a number of knowledge-based systems for interpreting IR spectral data. The development of each system has been found to assist with the development of later systems. Work continues to improve the performance of these systems by advancing the logic, knowledge, and underlying philosophical basis of these systems. The ultimate goal is to produce systems which assist, not replace, the chemist using them.

Research in our laboratories continues to explore the limitations of using a peak-based system for interpreting IR spectra of vapor-phase mixtures. We have concentrated on developing a version MIXIR for this purpose. Though the goal of this work is to improve the interpretation of IR spectra of vapor-phase species, it is expected that continued development of the logical paradigms incorporated into MIXIR will eventually improve condensed-phase analysis as well.

APPENDIXES

APPENDIX 8.1. PAIRS Decision Trace for the Functionality "AROMATIC" during the Interpretation of the Spectral Data Given in Table 8.1

Functionality AROMATIC	Passed initial empirical formula test

PEAK QUERY
    ANY PEAK(S)   POSITION: 2990–3050   INTENSITY: 1–10
                                        WIDTH: SHARP TO BROAD
    ANSWER——YES——
ACTION——SET AROMATIC   TO 0.20   CURRENT VALUE = 0.20

PEAK QUERY
    ANY PEAK(S)   POSITION: 1571–1620   INTENSITY: 1–10
                                        WIDTH: SHARP TO BROAD
    ANSWER——YES——
ACTION——ADD 0.15 TO AROMATIC   CURRENT VALUE = 0.35

PEAK QUERY
    ANY PEAK(S)   POSITION: 1571–1610   INTENSITY: 1–10
                                        WIDTH: SHARP
    ANSWER——NO——

PEAK QUERY
    ANY PEAK(S)   POSITION: 1571–1620   INTENSITY: 1–10
                                        WIDTH: SHARP
    ANSWER——NO——

PEAK QUERY
    ANY PEAK(S)   POSITION: 1571–1610   INTENSITY: 1–10
                                        WIDTH: SHARP TO BROAD
    ANSWER——YES——
ACTION——ADD 0.05 TO AROMATIC   CURRENT VALUE = 0.40

PEAK QUERY
    ANY PEAK(S)   POSITION: 1460–1520   INTENSITY: 1–10
                                        WIDTH: SHARP TO BROAD
    ANSWER——YES——
ACTION——ADD 0.10 TO AROMATIC   CURRENT VALUE = 0.50

PEAK QUERY
    ANY PEAK(S)   POSITION: 1481–1510   INTENSITY: 1–10
                                        WIDTH: SHARP
    ANSWER——NO——

PEAK QUERY
    ANY PEAK(S)   POSITION: 1481–1520   INTENSITY: 1–10
                                        WIDTH: SHARP
    ANSWER——NO——

Functionality AROMATIC	Passed initial empirical formula test

PEAK QUERY
    ANY PEAK(S)   POSITION: 1481–1510   INTENSITY: 1–10
                               WIDTH: SHARP TO BROAD
    ANSWER——YES——
ACTION——ADD 0.05 TO AROMATIC   CURRENT VALUE = 0.55

EXPECTATION QUERY
    IS AROMATIC   LESS THAN   0.200?
ANSWER——NO——

PEAK QUERY
    AT LEAST 2 PEAK(S)   POSITION: 1660–2000   INTENSITY: 1–3
                               WIDTH: SHARP TO AVERAGE
    ANSWER——YES——
ACTION——ADD 0.20 TO AROMATIC   CURRENT VALUE = 0.75

PEAK QUERY
    ANY PEAK(S)   POSITION: 660–900   INTENSITY: 2–10
                               WIDTH: SHARP TO BROAD
    ANSWER——YES——
ACTION——ADD 0.10 TO AROMATIC   CURRENT VALUE = 0.85

PEAK QUERY
    ANY PEAK(S)   POSITION: 660–900   INTENSITY: 7–10
                               WIDTH: SHARP TO BROAD
    ANSWER——YES——
ACTION——ADD 0.10 TO AROMATIC   CURRENT VALUE = 0.95

APPENDIX 8.2. PAIRS Decision Trace Including Explanatory Comments for the Functionality "AROMATIC-MONOSUBSTITUTED" during the Interpretation of the Spectral Data Given in Table 8.1

Functionality AROM-MONOSUBST	Passed initial empirical formula test

* (The following sample matrix is opaque in the definitive regions for Aromatic substitution class determination. If the unknown is run in this matrix, then, no information on these classes may be derived:)
QUERY—IS THE SOLVENT/SAMPLE STATE CHCL3?
    ANSWER——NO——

Functionality AROM-MONOSUBST   Passed initial empirical formula test

* (As before:)
QUERY—IS THE SOLVENT/SAMPLE STATE CCL4?
    ANSWER——NO——
* (The parent class must have some finite expectation, to justify investigation of the
    subclasses:)
EXPECTATION QUERY
    IS AROMATIC   GREATER THAN   0.000?
    ANSWER——YES——
ACTION——SET Z1   TO 0.950   CURRENT VALUE = 0.950
ACTION——MULTIPLY Z1   BY 0.100   CURRENT VALUE = 0.095
* (The increment of expectation to be applied to the subclass is scaled to the expectation of
    the parent class. This increment is Z1:)
* (Phenyl substitution classes may be differentiated by the characteristic pattern of the $C-H$
    out-of-plane bending absorptions observed in the 900 to 670 cm^{-1} region. This pattern
    reflects, among other things, symmetry-based vibrational degeneracy arising from the
    substitution class.
* The following queries seek to examine this pattern match, just as is done for the other
    substitution classes:)

PEAK QUERY
    ANY PEAK(S)   POSITION: 721–820   INTENSITY: 4–10
                         WIDTH: SHARP TO BROAD
    ANSWER——YES——
ACTION——SET AROM-MONOSUBST   TO 0.095   CURRENT VALUE = 0.095

PEAK QUERY
    ANY PEAK(S)   POSITION: 660–720   INTENSITY: 4–10
                         WIDTH: SHARP TO BROAD
    ANSWER——YES——
ACTION——AND 0.095 TO AROM-MONOSUBST   CURRENT VALUE = 0.190

PEAK QUERY
    ANY PEAK(S)   POSITION: 660–720   INTENSITY: 6–10
                         WIDTH: SHARP TO BROAD
    ANSWER——YES——
ACTION——ADD 0.095 TO AROM-MONOSUBST   CURRENT VALUE = 0.285

PEAK QUERY
    ANY PEAK(S)   POSITION: 721–820   INTENSITY: 6–10
                         WIDTH: SHARP TO BROAD
    ANSWER——NO——

APPENDIX 8.3. Trace of a User Interactive PAIRS Interpretation Session for the Functionality "AROMATIC-MONOSUBSTITUTED" during the Interpretation of the Spectral Data Given in Table 8.1 (The response of the user is *underlined*)

Functionality AROM-MONOSUBST   Passed initial empirical formula test

---

* (The following sample matrix is opaque in the definitive regions for Aromatic substitution class determination. If the unknown is run in this matrix, then, no information on these classes may be derived:)

QUERY—IS THE SOLVENT/SAMPLE STATE CHCL3?
    ANSWER——NO——
DESIRED ANSWER (YES, NO, RETURN): <u>NO</u>
THE ANSWER IS: NO
* (As before:)
QUERY—IS THE SOLVENT/SAMPLE STATE CCL4?
    ANSWER——NO——
DESIRED ANSWER (YES, NO, RETURN): <u>NO</u>
THE ANSWER IS: NO
* (The parent class must have some finite expectation, to justify investigation of the subclasses:)
EXPECTATION QUERY
    IS AROMATIC   GREATER THAN   0.000?
    ANSWER——YES——
DESIRED ANSWER (YES, NO, RETURN): <u>YES</u>
THE ANSWER IS: YES
ACTION——SET Z1   TO 0.0950   CURRENT VALUE = 0.950
ACTION——MULTIPLY Z1   BY 0.100   CURRENT VALUE = 0.0950
* (The increment of expectation to be applied to the subclass is scaled to the expectation of the parent class. This increment is Z1:)
* (Phenyl substitution classes may be differentiated by the characteristic pattern of the $C-H$ out-of-plane bending absorptions observed in the 900 to 670 cm^{-1} region. This pattern reflects, among other things, symmetry-based vibrational degeneracy arising from the substitution class.
* The following queries seek to examine this pattern match, just as is done for the other substitution classes:)

PEAK QUERY
    ANY PEAK(S)   POSITION: 721–820   INTENSITY: 4–10
                        WIDTH: SHARP TO BROAD

    ANSWER——YES——
DESIRED ANSWER (YES, NO, RETURN): <u>YES</u>
THE ANSWER IS: YES
ACTION——SET AROM-MONOSUBST   TO 0.095   CURRENT VALUE = 0.095
PEAK QUERY
    ANY PEAK(S)   POSITION: 660–720   INTENSITY: 4–10
                        WIDTH: SHARP TO BROAD

    ANSWER——YES——
DESIRED ANSWER (YES, NO, RETURN): <u>YES</u>
THE ANSWER IS: YES
ACTION——ADD 0.095 TO AROM-MONOSUBST   CURRENT VALUE = 0.190

Functionality AROM-MONOSUBST   Passed initial empirical formula test

---

PEAK QUERY
    ANY PEAK(S)   POSITION: 660–720   INTENSITY: 6–10
                                WIDTH: SHARP TO BROAD

    ANSWER——YES——
DESIRED ANSWER (YES, NO, RETURN): YES
THE ANSWER IS: YES
ACTION——ADD 0.095 TO AROM-MONOSUBST   CURRENT VALUE = 0.285

PEAK QUERY
    ANY PEAK(S)   POSITION: 721–820   INTENSITY:6–10
                                WIDTH: SHARP TO BROAD

    ANSWER——NO——
DESIRED ANSWER (YES, NO, RETURN): YES
THE ANSWER IS: YES
ACTION——ADD 0.095 TO AROM-MONOSUBST   CURRENT VALUE = 0.380

PEAK QUERY
    ANY PEAK(S)   POSITION: 730–770   INTENSITY: 6–10
                                WIDTH: SHARP TO BROAD

    ANSWER——NO——
DESIRED ANSWER (YES, NO, RETURN): YES
THE ANSWER IS: YES
ACTION——ADD 0.095 TO AROM-MONOSUBST   CURRENT VALUE = 0.475

PEAK QUERY
    ANY PEAK(S)   POSITION:730–770   INTENSITY: 7–10
                                WIDTH: SHARP TO BROAD
    ANSWER——NO——
DESIRED ANSWER (YES, NO, RETURN): NO
THE ANSWER IS: NO

PEAK QUERY
    ANY PEAK(S)   POSITION: 690–710   INTENSITY: 6–10
                                WIDTH: SHARP TO BROAD
    ANSWER——YES——
DESIRED ANSWER (YES, NO, RETURN): YES
THE ANSWER IS: YES
ACTION——ADD 0.095 TO AROM-MONOSUBST   CURRENT VALUE = 0.570

PEAK QUERY
    ANY PEAK(S)   POSITION: 690–710   INTENSITY: 7–10
                                WIDTH: SHARP TO BROAD
    ANSWER——YES——
DESIRED ANSWER (YES, NO, RETURN): YES
THE ANSWER IS: YES
ACTION——ADD 0.095 TO AROM-MONOSUBST   CURRENT VALUE = 0.570

Functionality AROM-MONOSUBST	Passed initial empirical formula test

PEAK QUERY
    ANY PEAK(S)   POSITION: 690–710   INTENSITY: 7–10
                             WIDTH: SHARP TO BROAD
    ANSWER——YES——
DESIRED ANSWER (YES, NO, RETURN): <u>YES</u>
THE ANSWER IS: YES
ACTION——ADD 0.095 TO AROM-MONOSUBST   CURRENT VALUE = 0.665

PEAK QUERY
    AT LEAST 4 PEAK(S)   POSITION: 660–900   INTENSITY: 6–10
                             WIDTH: SHARP TO BROAD
    ANSWER——NO——
DESIRED ANSWER (YES, NO, RETURN): <u>NO</u>
THE ANSWER IS: NO
ACTION—ADD 0.095 TO AROM-MONOSUBST   CURRENT VALUE = 0.760

PEAK QUERY
    2 PEAK(S)   POSITION: 660–900   INTENSITY: 6–10
                             WIDTH: SHARP TO BROAD
    ANSWER——NO——
DESIRED ANSWER (YES, NO, RETURN): <u>YES</u>
THE ANSWER IS: YES
ACTION——ADD 0.095 TO AROM-MONOSUBST   CURRENT VALUE = 0.855

REFERENCES

1. H. B. Woodruff and G. M. Smith, "Computer Program for the Analysis of Infrared Spectra," *Anal. Chem.* **52**, 2321–2327 (1980).
2. H. B. Woodruff and G. M. Smith, "Generating Rules for PAIRS-A Computerized Infrared Spectral Interpreter," *Anal. chim. Acta* **133**, 545–553 (1981).
3. G. M. Smith and H. B. Woodruff, "Development of a Computer Language and Compiler for Expressing the Rules of Infrared Spectral Interpretation," *J. Chem. Inf. Comput. Sci.* **24**, 33–39 (1984).
4. S. A. Tomellini, R. A. Hartwick, J. M. Stevenson, and H. B. Woodruff, "Automated Rule Generation for the Program for the Analysis of IR Spectra (PAIRS)," *Anal. Chim. Acta* **162**, 227–240 (1984).
5. S. A. Tomellini, R. A. Hartwick, and H. B. Woodruff, "Automatic Tracing and Presentation of Interpretation Rules Used by PAIRS: Program for the Analysis of IR Spectra," *Appl. Spectrosc.* **39**, 331–333 (1985).
6. B. J. Wythoff, C. F. Buck, and S. A. Tomellini, "Descriptive Interactive Computer-Assisted Interpretation of Infrared Spectra," *Anal. Chim. Acta* **217**, 203–216 (1989).
7. D. D. Saperstein, "Methodology for Evaluating and Optimizing Infrared Interpretations," *Appl. Spectrosc.* **40**, 344–348 (1986).
8. M. A. Puskar, S. P. Levine, and R. Turpin, "Compatibility Testing and Materials Handling," in: *Protecting Personnel at Hazardous Waste Sites* (S. P. Levine and W. F. Martin, eds.), Chapter 6, Butterworths, Woburn, MA (1985).

9. M. A. Puskar, S. P. Levine, and S. R. Lowry, "Computerized Infrared Spectral Identification of Compounds Frequently Found at Hazardous Waste Sites," *Anal. Chem.* **58**, 1156–1162 (1986).
10. M. A. Puskar, S. P. Levine, and S. R. Lowry, "Infrared Screening Technique for Automated Identification of Bulk Organic Mixtures," *Anal. Chem.* **58**, 1981–1989 (1986).
11. M. A. Puskar, S. P. Levine, and S. R. Lowry, "Qualitative Screening of Hazardous Waste Drum Mixtures," *Environ. Sci. Technol.* **21**, 90–96 (1987).
12. L. S. Ying, S. P. Levine, S. A. Tomellini, and S. R. Lowry, "Self-Training, Self-Optimizing Expert System for Interpretation of the Infrared Spectra of Environmental Mixtures," *Anal. Chem.* **59**, 2997–2203 (1987).
13. L. S. Ying, S. P. Levine, S. A. Tomellini, and S. R. Lowry, "Expert System for Interpretation of the Infrared Spectra of Environmental Mixtures," *Anal. Chim. Acta* **210**, 51–62 (1988).
14. B. J. Wythoff and S. A. Tomellini, "Generation of Compound-Specific Descriptions for Interpreting the Infrared Spectra of Condensed-Phase Mixtures," *Anal. Chim. Acta* **227**, 343–358 (1989).
15. B. J. Wythoff and S. A. Tomellini, "Dynamic, Computer-Assisted Interpretation of Infrared Spectra of Condensed-Phase Mixtures," *Anal. Chim. Acta* **227**, 359–377 (1989).
16. L. S. Ying and S. P. Levine, "Fourier Transform Infrared Least-Squares Methods for the Quantitative Analysis of Multicomponent Mixtures of Airborne Vapors of Industrial Hygiene Concern," *Anal. Chem.* **61**, 677–683 (1989).
17. H. K. Xiao, S. P. Levine, and J. B. D'Arcy, "Iterative Least-Squares Fit Procedures for the Identification of Organic Vapor Mixtures by Fourier Transform Infrared Spectrophotometry," *Anal. Chem.* **61**, 2708–2714 (1989).
18. B. J. Wythoff, S. P. Levine, and S. A. Tomellini, "Spectral Peak Verification and Recognition Using a Multilayered Neural Network," Anal. Chem. **62**, 2702–2709 (1990).

Fuzzy Rule-Building Expert Systems

Peter B. Harrington

9.1. INTRODUCTION

Modern trends in analytical spectroscopy have improved the performance of instrumentation. Performance may be defined by various measures. Spectrometers are continuously lowering their detection limits. Increased resolution and multidimensional experiments have enhanced the informing power furnished by spectra. Spatial and time resolved spectroscopy yield additional dimensions of information. Applications have grown with better and new instrument configurations. When applications require analyzing complex samples, the data may become both intricate and overwhelming.

Automated selection of relevant information from abundant data is an area of active research. This research is driven by several factors. There is a projected shortfall of analytical chemists. The demand for analytical information is increasing with growing environmental and litigious awareness in society. Improved automation and instrumentation will satisfy this demand with data, but the bottleneck occurs in the conversion of data to information.

Laser ionization mass spectrometry (LIMS) is an example of a novel experimental configuration that has opened up a range of applications. Semiconductor industries are interested in LIMS because small areas

Peter B. Harrington • Department of Chemistry, Clippinger Laboratories, Ohio University, Athens, Ohio 45701-2979.

Computer-Enhanced Analytical Spectroscopy, Volume 3, edited by Peter C. Jurs. Plenum Press, New York, 1992.

(2.5-μm diameter) of organic polymer thin films coated on integrated circuits may be directly sampled. LIMS spectra may appear to be different from conventional mass spectra. Laser ablation of a pure polymer may produce a complex mixture of photothermal products. Positive ion mode spectra may contain peaks at some unconventional masses (e.g., 1, 10, 11, 12, and 24) and other anomalies such as pseudoneutral losses of 11 and 12 mass units. Inorganic cations often appear in LIMS spectra. Ions with mass 10 and 11 are isotopes of boron. Sodium (23 m/z) and potassium (39 m/z) are also evident in the LIMS spectrum of Nylon 6 in Fig. 9.1. Spectral interpretation may be further complicated by matrix effects and inter-instrument variances.[1]

Interpretation of analytical data is typically accomplished by an expert. However, expertise for some applications may be unavailable, especially if the application is new and previous experience is not pertinent. For other cases, the expert may not be available or the expert may not be able to spend a sufficient amount of time on any given set of data. Rule-building expert systems are expedient because they can be used to generate expertise.[2–4] These systems store information in a rule format which consists of an antecedent (test for a condition) and a consequent (course of action). Rule-building expert systems are commercially available, but they generate rules which test only a single variable at a time.

A multivariate rule-building expert system (MuRES) was devised at the Colorado School of Mines as part of a pyrolysis-mass spectrometry data analysis package.[5] This technique has been applied successfully to other spectroscopic methods. This method is a useful pattern recognition method

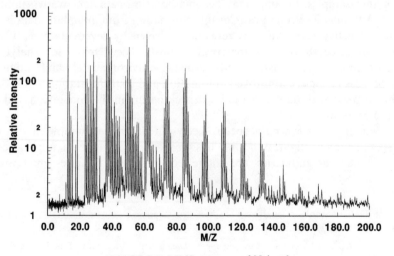

FIGURE 9.1. LIMS spectrum of Nylon 6.

that combines the benefits of linear discriminants and rule-building expert systems. However, MuRES has several limitations: it is computationally intensive, it is susceptible to spurious or ill-conditioned classification, it does not account for spectral distances in its induction, and it has difficulty with overlapping groups of data. A fuzzy rule-building expert system (FuRES) was devised to abate these difficulties. Unlike MuRES, FuRES generates its rules solely from fuzzy set and information theory. Using a fuzzy paradigm provides both efficient and effective optimization that has a simulated annealing aspect.[6]

9.2. MULTIVARIATE RULE-BUILDING EXPERT SYSTEMS

Classification of spectra has been studied since the late 1960s, when computers became accessible to chemists.[7,8] Previous research in the field of linear classifiers laid the foundations for the MuRES and FuRES methods.[9–13] Both MuRES and FuRES have a multivariate aspect in that their rules are linear discriminants. The incorporation of feature transformation provides a complementary alternative to univariate expert systems that rely on feature selection. A multivariate approach is advantageous for spectral data, because a spectrum often contains many redundant features and feature transformation in some cases may yield a signal-to-noise enhancement.

An abstract space may be defined by a set of spectra, for which each spectrum is represented as a vector whose components correspond to the spectral data points or channels. Many spectral classifiers partition the spectra space into subspaces where all the spectra have the same distinguishing property. The dimension of a partition is one less than that of the spectra space (i.e., number of variables). If a spectrum contains v data points, the spectra space is v-dimensional and the partitions will be $(v-1)$-dimensional hyperplanes. The hyperplane orientation can be defined by an orthogonal vector, which is referred to as a weight vector ($\mathbf{w}$). The location of the hyperplane can be defined by an attribute (b), which is the intersection of the weight vector and hyperplane. Figure 9.2a gives an example of a two-dimensional mass spectra space for the isomers of pentanol. The arrow indicates the weight vector for separating primary from secondary and tertiary pentanols. Figure 9.2b is a rendition of a fuzzy partition of the same space.

Projection of data objects onto a weight vector is useful, because the objects are mapped from a multidimensional space to a single dimension. The antecedent of the rule determines whether a projection onto the weight vector is greater or less than the attribute value. Selection of this point is

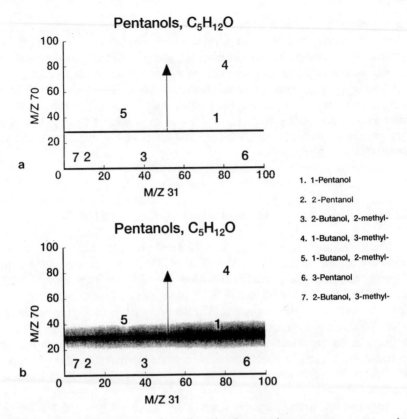

FIGURE 9.2. (a) Crisp partition of two-dimensional pentanol spectra space separates primary from secondary and tertiary alcohols. (b) Fuzzy partition of two-dimensional pentanol spectra space. The light band indicates the region where partitioning occurs to a small degree.

crucial to the classification efficacy and depends on the orientation of the hyperplane. Projections of data onto the weight vector are obtained by

$$x_k = \sum_{i=1}^{v} (w_i \times d_{k,i})/\|\mathbf{w}\| - b \qquad (9.1)$$

where x_k is the projected observation of d_k, v is the number of variables of the observation and of the weight vector, and b is the attribute value. The projection is corrected for the length of the weight vector. By subtracting b from the projection, the logic of the hyperplane separation is given by the sign of the projection.

Previous researchers have determined the attribute simultaneously with the hyperplane by adding an additional component of constant value to each spectrum.[11-13] Using an additional dimension to characterize the attribute may deleteriously affect optimization methods, because the attribute and the weight vector are dependent. In other words, any adjustment to the attribute position at a simplex vertex will affect the function values at the other vertices. Both MuRES and FuRES remedy this problem by determining the optimal attribute for each evaluation of a weight vector.

A minimum spanning tree of linear discriminants may be obtained by minimizing the entropy of classification, $H(C|A)$. Tree structure is determined by the iterative dichotimizer 3 (ID3) algorithm which is the basis for most rule-building expert systems.[14, 15] The algorithm constructs a minimal spanning tree of rules that minimize the entropy of classification, $H(C|A)$. The latter quantity is given by Eqs. (9.2) and (9.3):

$$H(C|a_j) = - \sum_{i=1}^{n} p(c_i|a_j) \ln p(c_i|a_j) \qquad (9.2)$$

$$H(C|A) = \sum_{j=1}^{m} p(a_j) H(C|a_j) \qquad (9.3)$$

where $H(C|a_j)$ is the entropy for a given attribute, a_j, which corresponds to less than or greater than zero; $H(C|a_j)$ is obtained by summing the entropy for each class, c_i, occurring for a_j, over the number of classes n. The probability, $p(c_i|a_j)$, is obtained by counting the number of observations of class i and dividing that number by the total number of observations with attribute j.

Equation (9.3) is the sum of the entropies for each attribute weighted by the prior probability that the attribute occurs. The number of attributes, m, is equal to 2 (the attributes are either positive or negative), and $p(a_j)$ is the number of observations with a given attribute divided by the total number of observations. Quantity $H(C|A)$ indicates more than the number of misclassifications of a rule; it also favors rule construction for which the misclassified points will have the least influence.

Several problems exist for the MuRES algorithm. The first problem is group overlap. Groups of spectra in space may be overlapping and provide ill-conditioned rule construction. A converse situation is when group underlap occurs. The groups are well separated and an infinite number of hyperplanes exist which all have the same $H(C|A)$. MuRES solved the problem of group underlap by maximizing the rule separation while maintaining a minimum entropy. After the rule is obtained, the critical value is centered at the median between the greatest negative projection and the least positive projection. For cases with outliers, centering the attribute in this manner may not be the best solution. Finding rules in a discrete spectra space also may be confounded by local entropy minima.

9.3. THE MATHEMATICS OF INEXACTNESS

Fuzzy set theory was developed by Zadeh in 1965 to accommodate inexact models of reality.[16] Although fuzzy logic has been slowly accepted in the United States, the Japanese have quickly adopted and implemented fuzzy logic in hardware. They have developed a variety of products which range from fuzzy elevators that give smooth rides to fuzzy vacuum cleaners that adjust the suction to the type and amount of dirt on the floor. Fuzzy logic has been applied to some problems in analytical chemistry.[17, 18]

A crisp set consists of objects that either belong or do not belong to the set. Fuzzy sets have elements which may partially belong. Therefore, the set boundary is not crisp, and the set appears blurred or fuzzy. Perhaps one of the hindrances to the acceptance of fuzzy techniques in the United States is the connotation that fuzziness obscures information and is the same as haziness. This assumption is false. Fuzzy representations can preserve information,[19, 20] because crisp representation is a subset of fuzzy representation.

Fuzzy set theory differs from classical set theory by some elementary operations. The intersection of two fuzzy sets is obtained from the minimum membership criteria for each member contained in both sets. The union of two sets is obtained from the maximum membership function of the two sets. Fuzziness is a measure of uncertainty in representation and differs from probability, which measures uncertainty in frequency of occurrence.*

---

* *Difference between Probability and Fuzziness.* Fuzzy set theory is often used by expert systems for assigning certainty factors to rules. For some cases there has been some confusion and interchanging of probability and fuzzy set membership values. However, fuzziness and probability are two distinctly different concepts.

The magnitude of the ion intensities at a given mass-to-charge ratio in a data set of LIMS spectra could be categorized as high or low. The fuzzy set of high intensities may be defined by the ratio of an ion intensity to the largest intensity in the spectrum. Therefore, if the intensity is the largest in a spectrum, it will have a value of fuzzy membership in the set of high intensities of 1.0, and if it is absent it will have a value of 0.0. The probability, for a given mass unit, may be defined by the frequency that a spectrum with a high intensity will be randomly selected from the training set. Case 1 has 0.1 intensities occurring for the complete set of spectra. Case 2 has intensities of 1.0 occurring for half the spectra. The table below demonstrates that probability and fuzziness each describe a different phenomenon.

	Uncertainty	
	Probability	Fuzziness
Case 1	1.0	0.1
Case 2	0.5	1.0

Some of the difficulties with the MuRES method have been alleviated by incorporating fuzzy set theory into the entropy calculation. Fuzzy set theory lends itself to inexact reasoning, and may be considered an infinite-valued logic system. Many concepts embodied in chemistry are inexact. Chemical compounds react to varying degrees and at varying rates. For example, when is a compound lipophilic or hydrophilic? Increasing the alkane portion of alkyl sulfonates gradually increases lipophilicity and decreases hydrophilicity. Alternatively, there are cases in chemistry for which set membership is crisp. For example, a molecule either has an ether group or it does not. FuRES does encompass these cases because it optimizes fuzziness within its rule construction.

Fuzzy rules are obtained by partitioning the spectra space into two fuzzy sets. For a rule that partitions data into two fuzzy sets, the set membership values may be used as logic for the rule. The logistic function is

$$\chi_A = 1.0/(1.0 + e^{-x/T}) \tag{9.4}$$

$$\chi_{A^C} = 1.0/(1.0 + e^{x/T}) \tag{9.5}$$

$$\chi_A + \chi_{A^C} = 1.0 \tag{9.6}$$

where χ_A is the degree of membership or belief in A and varies between 0 and 1, x is a variable that can vary between $-\infty$ and ∞, A^C is the degree that A does not occur (the converse), while T is a variable analogous to

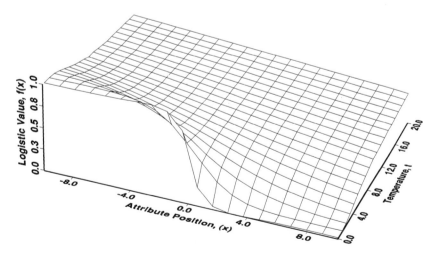

FIGURE 9.3. Plot of logistic values as a function of computational temperature and distance of a projection from the attribute: $f(x) = 1/(1 + e^{-x/t})$.

thermodynamic temperature and controls the degree of fuzziness. Henceforth, the variable T will be referred to as the computational temperature, which is given in the same units as the spectral intensities. We note that if the projections are not divided by the length of the weight vector in Eq. (9.1) t will become undefined. This logistic function is plotted in Fig. 9.3 as a function of temperature and distance from the attribute.

Although many logistic functions may be used, this one has some beneficial properties. The fuzziness itself may be controlled by T: as T approaches 0 the logistic function approaches the crisp case for which binary logic is obtained. Contrarily, as T gets large with respect to x, the logic will converge to a single value of 0.5 which is the maximum entropy case where no partitioning occurs. The function is differentiable and provides symmetric blurring of the hyperplane.

9.4. FUZZY RULE CONSTRUCTION

Fuzzy rule construction is similar to the method used by MuRES. The entropy of classification is the same, but the conditional probabilities $p(c_i|a_j)$ differ. Instead of counting discrete objects, they are counted by their logistic values:

$$p(c_i|A) = \sum_{k=1}^{n_i} \chi_A(x_k) \bigg/ \sum_{k=1}^{n} \chi_A(x_k) \tag{9.7}$$

Each x_k is an object projected onto the weight vector, the χ_A is the fuzzy set membership value of x_k in set A, and n_i is the number of objects in category i. A practical aspect of this fuzzy entropy calculation is that fuzzy entropy approaches crisp entropy as T approaches a value of zero.

Blurring the response surface of the entropy function is useful because it smooths out the local minima. The blurring occurs not for the data points, but for the hyperplane. The dimension of the hyperplane may be perceived as continuously decreasing from v to $v-1$ as a function of T and x given by

$$d = v - 1.0 + e^{-|x|/T} \tag{9.8}$$

where d is the dimensionality of the hyperplane, v is the dimensionality of the spectra space, x is the distance from the hyperplane, and T is the computational temperature. The total dimensionality of the space must be conserved. The dimensionality of the spectra space equals the dimension

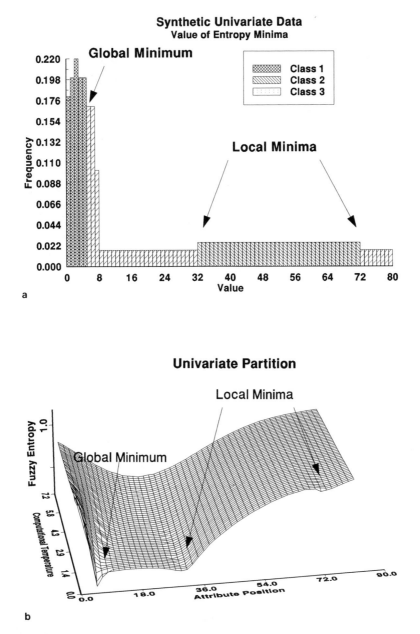

FIGURE 9.4. (a) Histogram of simulated data projections; the arrows indicate entropy minima. (b) Plot of fuzzy entropy as a function of computational temperature, and location of the attribute for simulated data projections in Fig. 9.3a.

of the weight vector plus the dimension of the hyperplane. For an object that exist on both sides of the hyperplane at once (i.e., $|x|/T \to 0$), the dimensionality of the hyperplane will equal v at this point in space. Therefore, the dimensionality of the weight vector should be zero. The dimension of the weight vector also varies as a function of the ratio of temperature and distance between a location in space and the hyperplane.

For example, suppose the projections of the spectra onto a vector are distributed as shown by the histogram in Fig. 9.4a. There are three entropy minima located between the class changes and indicated by the arrows. Figure 9.4b is a three-dimensional plot of the fuzzy entropy as a function of the attribute position and the computational temperature. At high temperatures only one minima exists. The global minimum entropy may be found by following the path of lowest fuzzy entropy on the attribute position and temperature response surface.

Unlike crisp entropy, fuzzy entropy values are continuous, so their derivatives exist. Derivative methods for optimization are very efficient and allow the number of function evaluations to be minimized. Minimizing the number of function evaluations makes locating the optimum attribute position within each function evaluation realizable. A tremendous efficiency is gained over the nonlinear simplex method, which uses many more function evaluations and must iteratively restart to verify that a solution is truly a global optimum. In addition, the logistic values are a function of the distance of the projections from the attribute. Consequently, the separation of spectra in space is implicit in the rule construction.

Finding the best attribute, b, is obtained by a line minimization that uses derivatives of fuzzy entropy of classification. The line is given by the projections on a weight vector. The method uses inverse parabolic interpolation of points and derivatives as described and given in Section 10.3 of Ref. 21. The brackets are obtained by using a previous solution and an incremental step. The bracketing algorithm is given in Section 10.1 of Ref. 21. The attribute value which furnishes the minimum entropy is obtained for each entropy evaluation using the line minimization algorithm.

Crisp classification trees often use a depth first search, for which a path defined by the rules is obtained to the category designation at the leaf of the tree. Fuzzy classification may require searching the entire tree, because the path may be blurred and no clear distinction may exist between true and false consequents of a rule.

A certainty factor indicates the degree of fit of an object to its class value. They are the product of logistic values from rules along the classification path (from the root to a leaf of the tree). Branches of the tree may be pruned early, when a logistic value below a preselected threshold is obtained. For leaf nodes of the same class designation, the certainty

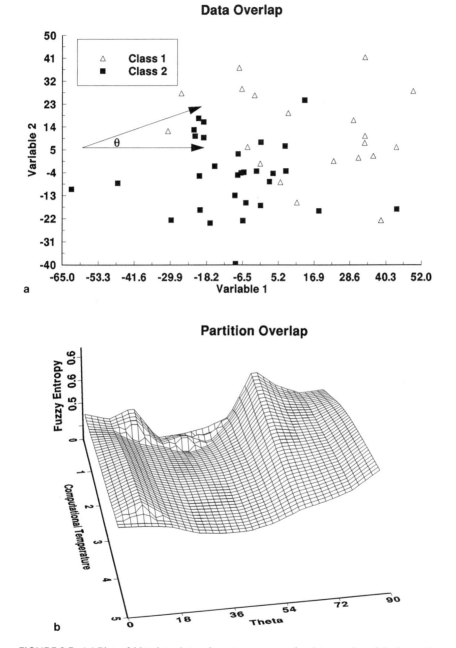

FIGURE 9.5. (a) Plot of bivariate data where two groups of points overlap; θ is the angle between the classifier and variable axes 1. (b) Plot of fuzzy entropy as a function of computational temperature and θ.

factors are added together, which provides AND/OR linkage of the rules. The certainty factors for an object will always sum to unity, because of the properties inherent in the logistic functions. Multiple solutions are supported with leaf nodes for different classes that have certainty factors greater than zero. The FuRES algorithm departs from fuzzy set theory with respect to its use of products and sums instead of using maximum and minimum logistic values to obtain certainty factors.

When the computational temperature is set too high, some of the projections may be located near the attribute. The OR logic of the fuzzy classification tree becomes important for unresolved points (i.e., close to the attribute).

The determination of the optimal computational temperature or degree of fuzziness is critical for rule generation. If the temperature is too high, a tree with many rules is obtained and the qualitative information is distributed among the rules. If the temperature is set too low, then local minima, overlap, and underlap become problems.

The FuRES algorithm uses a conjugate gradient method to find the weight vector with the lowest fuzzy entropy. The conjugate gradient algorithm is described and given in Section 10.6 of Ref. 21. This multidimensional algorithm consists of a sequence of line minimizations in conjugate directions (orthogonal to each other) until a minimum entropy is obtained. The line is determined by the gradient of fuzzy entropy, $\nabla H(C|\mathbf{w}, b)$. The attribute value with lowest entropy needs to be calculated one time for each gradient calculation.

The minimization starts at a high temperature to remove the local minima. A good starting temperature is one-fourth the furthest distance in spectra space between two objects of different class distinction. The classifier is refined iteratively at sequentially lower temperatures. The algorithm for this process is given in the Appendix. The fuzzy entropy will decrease as the temperature is lowered until either data overlap or underlap occurs. If the data are overlapped and not linearly separable, the fuzzy entropy will increase below some temperature which depends on the data. This case was simulated for two dimensions with normally distributed points about two group means. These points are shown in Fig. 9.5a. Figure 9.5b shows the fuzzy entropy as a function of temperature and θ, the angle of the classifier with respect to the first variable. If the data is underlapped, the entropy will not change. Plots similar to Figs. 9.5a and b are shown for underlapped data in Figs. 9.6a and b. If the algorithm terminates when the change in entropy, with respect to temperature, is less than or equal to zero, stable classifiers may be obtained.

FuRES uses a divide and conquer algorithm similar to other rule-building expert systems. The first rule will partition the training set of data into two subsets of data. The data are continually partitioned until each

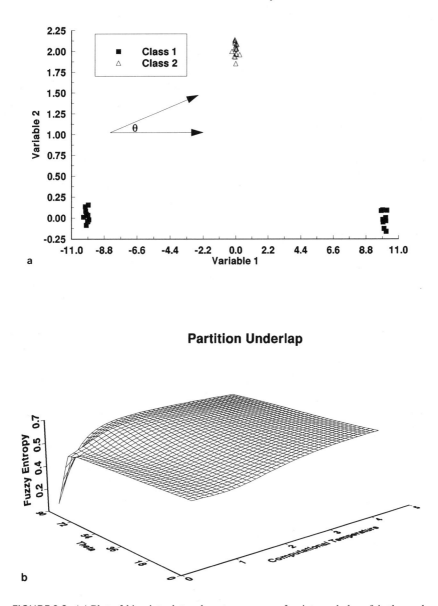

FIGURE 9.6. (a) Plot of bivariate data where two groups of points underlap; θ is the angle between the classifier and variable axes 1. (b) Plot of fuzzy entropy as a function of computational temperature and θ.

branch of the classification tree has a fuzzy entropy of zero. Degrees of freedom are lost, because the number of objects is decreased after each subsequent partition. However, because the fuzzy entropy expression seeks to maximize the distance of points from the hyperplane, it provides an impediment to overfitting.[22] A classification tree is constructed.

As mentioned earlier, the entire classification tree is searched with an unknown spectrum. The certainty factors for each rule are multiplied by those of the parent rule. A path down a branch of tree for which the certainty factor has reached a value of zero may be pruned from the search. Multiple conclusions are supported by certainty factors greater than zero. If several leaf nodes belong to the same category designations, their certainty factors are totaled. For any classification, only one certainty factor will be obtained per class and all the certainty factors will sum to unity.

9.5. CLASSIFICATION OF POLYMER THIN FILMS

The LIMS analyses were performed on the LIMA-2A laser microprobe instrument manufactured by Cambridge Mass Spectrometry, Ltd (Cambridge, England). The experimental configuration has been described

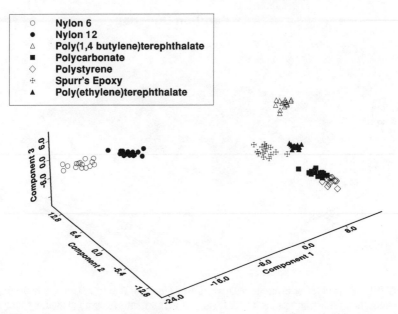

FIGURE 9.7. Plot of LIMS polymer scores on three linear discriminants which account for 97% of the total discriminant variance.

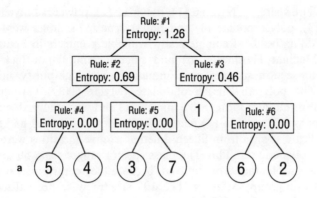

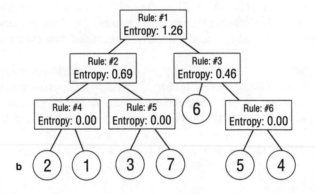

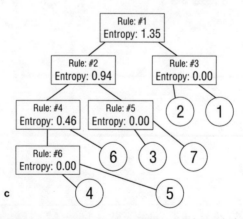

FIGURE 9.8. (a) MuRES classification tree; rectangles are rules and circles are leaf nodes which designate categories. (b) Degenerate MuRES classification tree, obtained from different simplex starting conditions. (c) FuRES classification tree.

in detail elsewhere.[23] Nylon 6 (1), nylon 12 (2), poly(1,4-butylene)terephthalate (3), polycarbonate (4) and polystyrene (5) samples were prepared by embedding beads of commercially available polymers in Spurr's Epoxy Resin (Medium Hardness) (6) and microforming 1-μm-thick sections. These thin-section samples were mounted onto high-purity silicon substrates. The poly(ethylene)terephthalate (Mylar 500D) (7) sample was provided in thin-film form ($\sim$0.5 mm thick) and was analyzed without any sample preparation. [(1)–(7) refer to leaf nodes in Figs. 9.8 and 9.9.]

Replicate spectra from fifteen different sample locations were obtained for each polymeric thin film. The mass spectra ranged from 50 to 300 amu. Masses below 50 amu were removed, because they contributed significantly to the within-group variances. The 105 spectra were normalized to unit vector length and then scaled by the average of the group standard deviations. The spectra were compressed by eigenvector projection to produce abstract spectra composed of 30 scores, which accounted for 98% of the total variance. The abstract spectra were used for MuRES and FuRES evaluation, because they furnish an overdetermined representation of the data.

The data matrix was a 105 by 30 matrix of single-precision floating-point numbers. Tree construction for MuRES took approximately 4 hours on an Everex Step 80486 computer operating at 25 MHz under DOS 4.01 using a cooling rate of 1.1. FuRES tree construction required less than 5 minutes on the same system. Figure 9.7 is a plot of the first two linear discriminant functions which account for 97% of the discriminant variance.

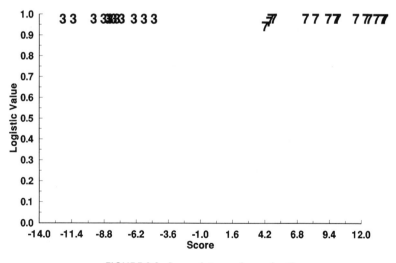

FIGURE 9.9. Score plot on rule number 5.

Figures 9.8a and b give two equivalent tree structures acquired from MuRES. These two trees were obtained by varying the initial positions of the simplex vertices. Because the data space is discrete, the simplex has a tendency not to explore other rules which are globally optimally. In other words, even with exhaustive iteration with randomized starting conditions and a continuous objective function, the simplex may not find a global

Poly(ethylene)terephthalate

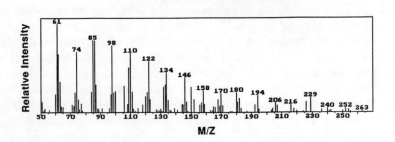

Rule Plot # 5

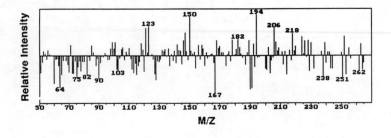

Poly(1,4-butylene)terephthalate

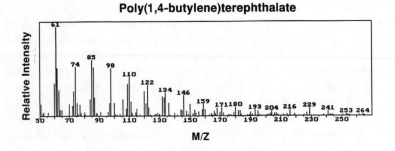

FIGURE 9.10. Fuzzy classifier for rule number 5.

optimum. The reason for this behavior may be caused by including the previous optimal weight vector as one of the simplex starting vertices. The simplex then has a vertex fixed and cannot move freely through the response surface. Eventually, the simplex may contract to give a slightly improved local optimum. A FuRES classification tree is given in Fig. 8C. FuRES is not as susceptible to choice of the initial vector, so long as the initial temperature is sufficiently large to give only one local minimum of the response surface. The MuRES tree structure does not indicate the larger distances between Nylon clusters (∘ and • in Fig. 9.7) and the other polymer clusters in spectra space. The structure of the FuRES tree characterizes this relationship.

Besides interpreting the qualitative structure of the classification tree, multivariate rule-building expert systems allow the weight vector loadings and the training data scores to be visualized. Both scores and loadings provide a great deal of qualitative information regarding the reliability of classification and the nature of the training set. Figure 9.9 is a score plot of the spectra onto rule 5. Large separations of scores imply a higher confidence in regard to the rule reliability. Figure 9.10 gives the weight vector for separating the polymers of terephthalate spectra. The loading of the weight vector indicates variables that are correlated and discriminatory within the context of the rule.

Internal validation, "the leave one out method" was used to compare FuRES classification with MuRES. The abstract spectra were not recalculated during this evaluation. The entire data set of spectra was correctly classified by FuRES, while linear discriminant analysis correctly classified 100 of the 105 spectra. More extensive evaluations are reported elsewhere.[22]

9.6. CONCLUSION

FuRES has the advantages of both efficiency and efficacy over the MuRES method. The FuRES method determines the optimal degree of fuzziness, and can accommodate crisp cases when warranted. FuRES inherently represents interclass distances in its rule construction. FuRES incorporates some modeling characteristics, because it uses distances in rule formation. As a result, FuRES provides more stable rule formation and avoids classification by small or spurious features in the spectra. This quality imparts robustness for nonlinearly separable cases. Although many similarities exist between FuRES and neural networks, FuRES has the advantage that its rules contain qualitative information which are amenable to visual interpretation.

APPENDIX. ALGORITHM FOR FINDING OPTIMUM t, $\mathbf{w}$, AND b

Initialization:

(1) $t_0 = \text{Max}[\|\mathbf{d}_1 - \mathbf{d}_2\|]/2$ Select initial temperature t_0 as the maximum distance between two observations $\mathbf{d}$ of different class designation.

(2) $\mathbf{w}_0 = \text{Max}[\mathbf{d}_1 - \mathbf{d}_2]/\|d_1 - d_2\|$ Select initial weight vector $\mathbf{w}_0$ as the difference normalized distance between two observations of different class distinction.

(3) $b_0 = \text{Min}[H(C|t_0, \mathbf{w}_0)]$ Obtain the best attribute, b_0 for t_0 and $\mathbf{w}_0$ using Line Minimization 10.3, Ref. 20.

Cooling:

(4) $\mathbf{w}_{n+1}, b_{n+1} = \text{Min}[H(C|t_n, \mathbf{w}_n, b_n)]$ Find best $\mathbf{w}_{n+1}$ and b_{n+1} is implicit by Congugate Gradient Minimization 10.6, Ref. 20.

(5) $t_{n+1} = t_n/g$ g is cooling rate, typical value range between 1.1 and 2.0.

(6) IF $(H_{n+1} < H_n)$ THEN $\mathbf{w}_n = \mathbf{w}_{n+1}$, $b_n = b_{n+1}$, $t_n = t_{n+1}$

(7) IF $(H(C|t_n, \mathbf{w}_n, b_n) - H(C|t_{n+1}, \mathbf{w}_{n+1}, b_{n+1}))/(t_n - t_{n+1}) > 0$
THEN GOTO 4.   Determine if the derivative is greater than zero goto step 4.

Entropy Calculation Internal in Gradient Calculation and Line Minimizations:

(7) $b_n = \text{Min}[H(C|\mathbf{w}_n, t_n)]$ For every entropy calculation find line minimum b_n by 10.3, Ref. 20.

(7A) $x_k = (\mathbf{d}_k^T \bullet \mathbf{w}_n)/\|\mathbf{w}_n\| - b_n$ Project data onto weight vector and subtract attribute value. Weight vector must be normalized, so that t_n is constant.
    Eq. (9.1)

(7B) $\chi_A(x_k) = 1.0/(1.0 + e^{-1.0 \times x_k/t_n})$

(7C) $p(c_i|A) = \sum_{k=1}^{n_i} \chi_A(x_k) \bigg/ \sum_{k=1}^{n} \chi_A(x_k)$ Eq. (9.7)

$$\text{(7D)} \quad H(C \mid a_j) = - \sum_{i=1}^{n} p(c_i \mid a_j) \ln p(c_i \mid a_j) \qquad \text{Eq. (9.2)}$$

$$\text{(7E)} \quad H(C \mid t_n, \mathbf{w}_k, b_n) = \sum_{j=1}^{m} p(a_j) H(C \mid a_j) \qquad \text{Eq. (9.3)}$$

Partitioning:

(8) If $\chi_A < K$ omit from partition A. K determines the rule redundancy
If $\chi_A < K$ omit from partition A^C.   Typical value is 0.5.

(9) Continue partitioning until $H(C \mid t_n, \mathbf{w}_n, b_n) = 0$.

ACKNOWLEDGMENT

This research was supported by Teledyne CME, Ohio University, and National Biscuit Company. Charles Evans and Associates, Filippo Radicati di Brozolo, and Robert W. Odom are thanked for collecting and supplying the LIMS polymer data. Busolo Wabuyele, Alan Hendricker, Peter Tandler, Peng Zheng, and Peter Rothman are thanked for their helpful comments and criticisms.

REFERENCES

1. K. J. Voorhees, E. W. Sarver, P. B. Harrington, and S. J. DeLuca, *Appl. Spectrosc.*, **45**, 36 (1991).
2. W. A. Schlieper, J. C. Marshall, and T. L. Isenhour, *J. Chem. Inf. Comput. Sci.* **28**, 159 (1988).
3. M. P. Derde, L. Buydens, C. Guns, D. L. Massart, and P. K. Hopke, *Anal. Chem.* **59**, 1868 (1987).
4. P. B. Harrington, K. J. Voorhees, T. E. Street, F. R. di Brozolo, and R. W. Odom, *Anal. Chem.* **61**, 715 (1989).
5. P. B. Harrington and K. V. Voorhees, *Anal. Chem.* **62**, 729 (1990).
6. J. H. Kalivas, N. Roberts, and J. M. Sutter, *Anal. Chem.* **61**, 2024 (1990).
7. P. C. Jurs and T. L. Isenhour, *Chemical Applications of Pattern Recognition*, Wiley, New York (1975).
8. N. A. B. Gray, *Computer Assisted Structure Elucidation*, Chapter 5, Wiley, New York (1986).
9. P. C. Jurs, B. R. Kowalski, and T. L. Isenhour, *Anal. Chem.* **41**, 21 (1969).
10. P. C. Jurs, B. R. Kowalski, T. L. Isenhour, and C. N. Reilley, *Anal. Chem.* **41**, 695 (1969).
11. G. L. Ritter, S. L. Lowry, C. L. Wilkins, and T. L. Isenhour, *Anal. Chem.* **47**, 1951 (1975).
12. T. R. Brunner, C. L. Wilkins, T. F. Lam, L. J. Soltzberg, and S. L. Kaberline, *Anal. Chem.* **48**(8), 1146 (1976).
13. S. L. Kaberline and C. L. Wilkins, *Anal. Chem. Acta* **103**, 417 (1978).
14. J. R. Quinlan, in: *Machine Learning: An Artificial Intelligence Approach* (R. S. Michalski, J. G. Carbonell, and T. M. Mitchell, eds.), p. 463, Tioga Publ. Co., Palo Alto, CA (1983).

15. B. Thompson and W. Thompson, *Byte* **11**, 149 (1986).
16. L. A. Zadeh, *Inf. Control* **8**, 338 (1965).
17. O. Mathias, *Anal. Chem.* **62**(14), 797A (1990).
18. O. Mathias, *Chemometrics and Intelligent Laboratory Systems* **4**, 101 (1988).
19. B. Kosko, *Inf. Sci. (N.Y.)* **40**, 165 (1986).
20. B. Kosko, *IEEE Trans. Pattern Analysis and Machine Intelligence* **PAMI-8**, 556 (1986).
21. W. H. Press, B. P. Flannery, S. A. Teukolsky, and W. T. Vetterling, *Numerical Recipes in C*, Chapter 10, Cambridge University Press, Cambridge (1988).
22. P. B. Harrington, *J. Chemometrics* (in press).
23. T. Dingle, B. W. Griffiths, J. C. Ruckman, and C. A. Evans, Jr., *Microbeam Analysis—1982* (K. F. J. Heinrich, ed.), p. 365, San Francisco Press, San Francisco, CA (1982).

Advanced Signal Processing and Data Analysis Techniques for Ion Mobility Spectrometry

10

Dennis M. Davis and Robert T. Kroutil

10.1. INTRODUCTION

Ion mobility spectrometry (IMS) is based on the drift of molecular ions through a gas of uniform temperature and pressure, in the presence of a weak electric field. When the electric field is applied uniformly throughout the gas, the ions begin to move along the field lines. This movement continues until the ions are retarded by collisions with the neutral gas molecules. Because the electric field is still present, the ions are accelerated once again and the entire process is repeated (this is in contrast to mass spectrometry, where there is no collision process). An average velocity is determined by millions of these accelerations and energy-losing collisions of the ions. The time required for the ion to traverse a region of known length is the drift time. Plotting the number of ions versus time results in the ion mobility spectrum.

Dennis M. Davis and Robert T. Kroutil • Analytical Division, Research Directorate, U.S. Army Chemical Research, Development, and Engineering Center, Aberdeen Proving Ground, Maryland 21010-5423.

Computer-Enhanced Analytical Spectroscopy, Volume 3, edited by Peter C. Jurs. Plenum Press, New York, 1992.

The average velocity of the ions, or drift velocity, is related to the strength of the electric field through the equation

$$v_d = l_d/t_d = KE \qquad (10.1)$$

where v_d is the drift velocity, l_d the distance over which the ions drift, t_d the drift time (the time it takes the ions to drift over the distance l_d), and E is the electric field strength; the constant of proportionality, K, is the "mobility" of the ions and is dependent upon both the ions and the neutral gas through which the ions move. The mobility of the ions at standard temperature and pressure, the "reduced mobility" K_0, is related to the mobility of the ions through the equation

$$K_0 = K[273/T][P/760] \qquad (10.2)$$

where T is the absolute temperature of the drift region and P is the total pressure of the gas and ions in the drift region. In this study, there is no need to correct the spectra for temperature and pressure since each spectrum is normalized with respect to the drift time.

The equation for the mobility of an ion through a gas, as given by Mason et al.,[1, 2] has been shown to be dependent on the first-order collision integral. The collision integral is proportional to the transport cross section, and thus the mobility and the drift velocity of the ions are dependent on the size of the ions, and also on the shape and distribution of the charge of the ions. More detailed discussions of ion mobility can be found elsewhere,[3] and the utility of IMS as an analytical tool for the rapid detection of airborne vapors in the atmosphere has been demonstrated.[4–7]

In the ion mobility spectrometer, the sample is introduced via an inlet probe. A portion of the sample crosses a semipermeable membrane into the ionization chamber, while the remainder of the sample is vented through the exhaust to the pumps. The carrier flow gas, clean air, is input directly into the ionization chamber. The sample ions and air ions are then ionized in the chamber and allowed to mix and interact. A driving pulse of known shape and duration is then applied to the gating grid to allow the mixture to enter the drift region of the spectrometer. The sample ions and air ions are then accelerated through the electric field in the drift region until they are retarded by collisions with neutral gas molecules. The ions then strike the collector electrode, after which the signal is processed to produce the ion mobility spectrum. A schematic diagram of an ion mobility spectrometer is shown in Fig. 10.1.

The past several years have seen the advance of IMS as an analytical technique. Most of these advances have been made in the hardware development end of the problem, with the result that portable IMS devices

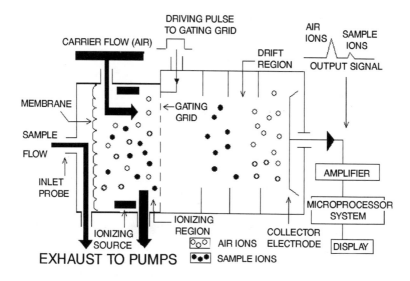

FIGURE 10.1. Schematic diagram of an ion mobility spectrometer.

have begun to appear in the marketplace. The other end of the problem, the signal processing and data analysis techniques, has not been addressed to the same degree. Recent attempts at applying data analysis techniques to IMS data have been made, and the results are encouraging. Data processing algorithms, ranging from those which perform simple tasks to those performing more difficult tasks, have been developed. Among the algorithms to be discussed are digital filtering for detecting peaks in the presence of interferents, and linear discriminant analysis algorithms for detecting and identifying industrial chemicals at or near their maximum exposure limits.

10.2. EXPERIMENTAL

10.2.1. Equipment

All data used in this study were collected on a hand-held IMS spectrometer, the Airborne Vapor Monitor (AVM) from Graseby Analytical Ltd., Watford, Herts, United Kingdom. The data for the digital filtering work was stored on a Nicolet Model 4094 oscilloscope. Each spectrum normally consisted of 3968 data points, which were collected at 5-microsecond intervals. The spectra for the linear discriminant analysis was stored on a Zenith personal computer, using a Graseby Ionics, Ltd. Advanced Signal Processing (ASP) board and its associated software. All of the

TABLE 10.1. Operational Parameters for the AVM

Number of waveforms to be summed	32
Number of samples per waveform	640
Gating pulse repetition rate	40 Hz
Gating pulse width	180 μs
Delay to start of sampling	0 μs
Sampling frequency	30 kHz
Grating pulse source is	external

spectra for the linear discriminant work was collected with the parameters shown in Table 10.1.

10.2.2. Pre-Processing of Spectra for Digital Filtering

The data for the digital filtering work were stored on a Nicolet oscilloscope and normally had 3968 data points. A data reduction procedure was performed to decrease the number of data points in the spectrum from 3968 data points to 512. The reduced number of data points was still twice the number used by the conventional IMS signal processing algorithm used in the AVM. This increase in the number of data points over that used in the hand-held IMS algorithm simply gives greater resolution in the processed spectra. However, because the spectral features in the IMS data are broad, no new spectral features are revealed by the increased resolution of the data file. The number of data points was selected as a power of two so that many of the computer programs already developed (such as Fourier transforms, etc.) could be used on the IMS data. In addition to the deresolution of the Nicolet data, the spectrum was normalized with respect to the X-axis. To do this, each value on the X-axis was divided by the value position of a reactant ion peak. For negative ion spectra, the peak position with the maximum intensity between 6.0 and 7.0 milliseconds drift time was used for the identification of the reference ion peak. For positive ion data, a value between 6.5 and 7.5 millisecond drift time was used as a window in which to find the reference ion peak. This new spectrum also appears as a pseudo-"Reduced Mobility" spectrum, which has a dimensionless X-axis corresponding to a ratio of drift times, T_R. Only the data in the range 0.5 to 3.0 along the T_R-axis are used. The IMS data files used in this study have data points every 0.05 T_R.

10.2.3. Pre-Processing of Spectra for Linear Discriminant Analysis

The pre-processing and data processing procedure used in the linear discriminant analysis is shown in Fig. 10.2. The first pre-processing step is

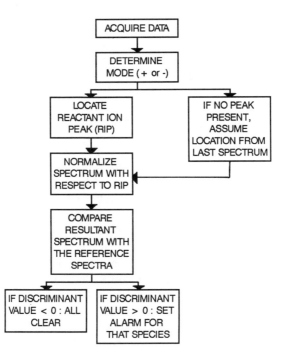

FIGURE 10.2. Block diagram showing the steps taken when performing a linear discriminant analysis on the ion mobility spectra.

to determine if the spectrum has been collected in the positive (+) or negative (−) mode. This knowledge is important since the Graseby ASP board does not differentiate between the two types of spectra, i.e., the ASP board converts all spectra to positive values. A preliminary discrimination is made based on the mode; a spectrum collected in the negative mode has no chemical semblance to a spectrum collected in the positive mode. Once the mode has been determined, it is necessary to determine the time at which the reactant ion peak (RIP) appears. The reactant ion for the AVM, O_2^- in the negative mode and H_3O^+ in the positive mode, is the species which transfers the charge to the chemical species being analyzed. The location of the RIP must be determined for each spectrum, if possible, because the location is affected by changes in temperature, pressure, and relative humidity. If no RIP is found, then one must assume the RIP is located at the same time as the RIP for the previous spectrum. After determining the time at which the RIP appears, the spectrum is normalized as described above to create a dimensionless X-axis.

10.3. DIGITAL FILTERING

Conventional IMS data processing algorithms often rely on baseline correction techniques to correct for baseline slope before peak picking routines are used. Problems which result from the baseline correction are signal suppression, or the failure to resolve overlapping peaks. An example of the failure to resolve overlapping peaks is encountered in the ion mobility spectra obtained when a challenge of bis-2-chloroethyl sulfide, phenol, or a mixture of bis-2-chloroethyl sulfide and phenol is introduced into an ion mobility spectrometer. It has been noted that the IMS peaks of phenol are much more intense than those of bis-2-chloroethyl sulfide for equal concentrations of the two species and the mobilities of the phenol peaks and the bis-2-chloroethyl sulfide peaks are not vastly different. This similarity of mobilities can create a problem for IMS peak detection and identification algorithms which are commonly used. The data set used in this study comprised IMS data collected on a hand-held atmospheric vapor monitor which uses a "conventional" peak detection and identification algorithm. The IMS detection algorithm used in the hand-held monitor utilizes a baseline correction routine that uses the average intensity of points to either side of the peak to be detected for determining the baseline. For the case of a mixture of phenol and bis-2-chloroethyl sulfide, this procedure breaks down. This breakdown is due to the fact that one of the points for determining the baseline falls near the top of the phenol peak, while the other point falls on the actual baseline. This results in a determined baseline which has a greater amplitude than the actual baseline; in many cases, the bis-2-chloroethyl sulfide peak intensity falls below this baseline level and the peak is ignored. Thus, an effort has been made to apply novel signal processing techniques, in particular digital filters, to the analysis of IMS spectral signatures to eliminate the baseline correction procedure. Several signal processing routines developed for infrared spectroscopy[8] have been adapted for the discrimination of IMS data.

10.3.1. Development of Digital Filters

It was decided to limit the processing techniques to those methods which were not computer intensive (i.e., those which do not require a large amount of computer processing time). Among the techniques that fit this criterion are curve fitting, spectral deconvolution, and digital filtering. The application of curve-fitting routines involves fitting a polynomial expression to the background spectral features, and subtracting this expression from the overall spectrum. Typically, this procedure does not work very well, as the background features are difficult to fit with a polynomial expression. In addition, for the case of IMS spectra, the spectral features can change as a result of changing the concentration of the vapor challenge. A better approach is the technique of spectral deconvolution. In spectral deconvolution, an inverse Fourier transform would be applied to a raw IMS spectrum to create an interferogram. Subsequently, apodization functions that use the spectral features to select the bands of interest are applied to the interferogram. A third technique, which is similar to spectral deconvolution, is digital filtering. Digital filtering has two advantages over spectral deconvolution in that (1) digital filters may be applied to either interferograms or time domain spectra, and (2) the spectral cutoff frequencies can be optimized for a given set of conditions. Like spectral deconvolution, the digital filters to be applied in the interferogram can be Fourier transformed and applied to the raw spectrum. For this work, digital filtering of the interferogram was chosen as the signal processing technique because of its advantages over the other techniques mentioned above.

Digital filters in the interferogram domain are functions in which a data transformation has been made to a new set of coordinates. In this case, the data transformation limits many of the background spectral features, while enhancing the spectral features associated with the bandpass of the digital filter. In this manner, a digital filtering technique can be shown to have a finite impulse response (FIR). That is, the filter is considered to be nonrecursive.[9] A nonrecursive digital filter has the functional form

$$Y_i^* = \sum_{i=1}^{n} f_i Y_{i-n} \tag{10.3}$$

where Y_i^* is the filtered data point, Y_i is a raw data point surrounding the original data point to be filtered, and f_i is the filter coefficient. Each term in the series $f_i Y_{i-n}$ is a weighted value that yields a different coefficient f_i for each data point. The summation of the filter terms yields a filtered point.

The type of filter shown in Eq. (10.3) can be defined in more general terms as a Z-transformation.[8] A digital filter is simply a Z-transformation in which a number of weighting coefficients are used. The Z-transform may be a complex function, but a linear function is generally used. An example of a Z-transform is given by the equation

$$X(z) = \sum_{n=0}^{\infty} x(n) z^{-n} \tag{10.4}$$

Using this form the value $X(z)$ can be defined as being equal to a power series of successive values of the time domain signal $x(n)$.

An all-zero model, where all of the Z-transform elements are in the numerator, was used on the IMS data. An all-pole model, where all of the Z-transform elements are in the denominator, or a model that has some zero terms and some pole terms, could also have been used. All-pole models are generally used for very sharp, narrow spectral features. The bandwidth of the features associated with the AVM data are relatively broad, resulting in the selection of the all-zero model. In addition, all-zero models are advantageous because they can have terms that lie anywhere in the Z-plane. For models that contain functions in the denominator, the transfer function of the poles should not lie outside of the "unit circle."[8] That is, the poles should not lie outside of the transfer function $(1 + Z^{-1})/(1 + Z^{-2})$. Instability of the digital filter will result if any of the poles are outside of the unit circle transfer function. For these reasons, an all-zero transfer function was used for the AVM data.

Infrared-spectroscopy-based interferogram software is being used, so the derivation of the filter coefficients must begin in the "drift time" domain. Effectively, a drift time filter is desired that possesses a bandpass centered on a characteristic spectral feature of the target species. An interferogram digital filter is equivalent to its frequency domain counterpart. The convolution of a time domain interferogram with a filter function is equivalent to the multiplication of a frequency response by an original IMS spectrum. The weighting coefficients can be obtained by relating the convolution of the frequency domain function with its corresponding unmodified time domain function. Thus, the selection of the proper weighting coefficients can essentially be reduced to a linear model that relates the filtered variables (the dependent variables) to the set of nonfiltered interferogram variables (the independent variables). The problem of generating the filter coefficients involves the use of a multiple "stepwise" regression procedure.[8] This procedure allows one to use a number of data points behind the point being filtered. Consecutive numbers of terms were added to the digital filter until the correlation coefficient did not vary with the addition of more terms.

A digital filter developed in this manner is a tailored procedure of selecting the proper weighting coefficients. The resulting digital filter should have a low gain at frequencies lower than the bandpass of the filter, a high gain at frequencies between the bandstops, and a low gain at frequencies higher than the bandpass. The filter developed by the above procedure is commonly termed a "notch" filter.

For the AVM and infrared interferograms, filtering points behind the point to be filtered were used. In this case, as shown in Eq. (10.3) and Reference 8, the modified equation can be written as

$$Y_i^* = f_i Y_{i-n} + \cdots + f_{n+1} Y_i \tag{10.5}$$

The filter developed for processing an IMS spectral feature consisted of a 25-term function, obtained using the raw datum point and the 24 previous data points.

10.3.2. Discussions

Before any advanced data processing was performed, it was important to know what information was available in the original spectra. Four spectra which are contained in a database are shown in Fig. 10.3. The spectra labeled A and B are the modified IMS spectra (i.e., those modified for the database as described above) which are obtained when the AVM is exposed to a vapor challenge containing 0.1 microgram/liter phenol in air. The AVM exhibited no response to the vapor challenge of the phenol at this concentration. No response means the AVM background compensation/peak detection algorithm was unable to identify the presence of a peak of interest. When only phenol is present, no peak could be found at $T_R = 1.36$, which is the ratioed drift time of the bis-2-chloroethyl sulfide peak. Spectra C and D are the modified IMS spectra obtained when the AVM is exposed to a vapor containing 0.1 microgram/liter phenol and 0.16

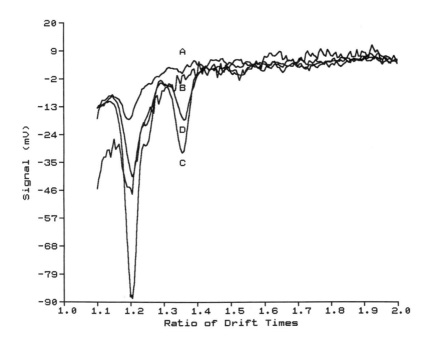

FIGURE 10.3. IMS spectra modified by correction of the reference ion O_2^-. Spectra A and B contain only phenol. Spectra C and D contain both phenol and bis-2-chloroethyl sulfide.

microgram/liter bis-2-chloroethyl sulfide in air. The AVM exhibited a
response to spectrum C and also a response to spectrum D.

In order to determine the effects that higher concentrations of phenol
and/or bis-2-chloroethyl sulfide have on the spectral features obtained on
the AVM, a number of other spectra were obtained. Figure 10.4 shows the
modified IMS spectra contained in database files labeled E, F, G, and H.
Spectra E and F are the spectra obtained when the AVM is exposed
to a vapor challenge containing 0.3 microgram/liter phenol and 0.16
microgram/liter bis-2-chloroethyl sulfide. The IMS did not exhibit any
responses to the vapor challenges. No response is given for this spectrum,
because the baseline correction algorithm did not indicate the presence of
a feature at a T_R of 1.36. This effect was due to the phenol peak located
in the range 1.2 to 1.3 being much more intense than the bis-2-chloroethyl
sulfide peak. When the baseline algorithm calculated the peak maximum of
the bis-2-chloroethyl sulfide feature, one end of the baseline is located on
the edge of the phenol peak. This resulted in a subtraction of a baseline
that was larger than the actual bis-2-chloroethyl sulfide peak. Spectrum G
is the modified IMS spectrum obtained when the AVM was exposed
to a vapor challenge containing 1.0 microgram/liter phenol and 0.16

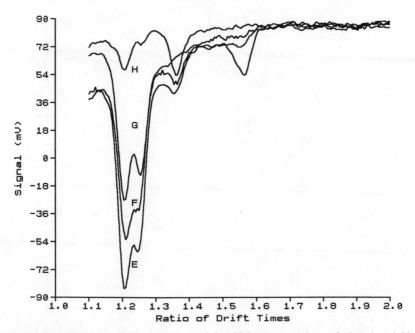

FIGURE 10.4. Corrected IMS spectra containing high concentrations of phenol and bis-
chloroethyl sulfide. Spectra E, F, and G contain a high concentration of the phenol spectral
interferant. Spectrum H contains a high concentration of bis-2-chloroethyl sulfide.

microgram/liter bis-2-chloroethyl sulfide. Again, the AVM exhibited no response to the vapor challenge. The last spectrum shown in Fig. 10.4 is spectrum H. This modified IMS spectrum was obtained when the AVM was exposed to a vapor containing 1.0 microgram/liter phenol and 0.4 microgram/liter bis-2-chloroethyl sulfide. The AVM had a response to the vapor challenge used for spectrum H. Analysis of the spectrum indicates that, although a response was given to the presence of bis-2-chloroethyl sulfide, the response was not caused by the peak at T_R 1.36, but was due to a peak at T_R 1.54, the location of the peak used by the AVM for the detection of another chemical species of interest. In this spectrum, the peak at T_R 1.54 is a phenol peak observed at high phenol concentrations.

In order to filter out the phenol spectral features for the case as shown in Fig. 10.3 (C and D) and Fig. 10.4 (E, F, and G), an Inverse Fast Fourier Transform (IFFT) has been applied to each spectrum. This procedure is performed, because the software used in this study has been adapted from infrared spectroscopic studies.[9, 10] Subsequently, a Fast Fourier Transform (FFT) is applied to each spectrum to re-create the initial "drift time" domain spectrum. These two manipulations of the original data were performed to aid in the visualization of whether any improvements can be made as a result of a linear digital filter. The resultant spectra are shown

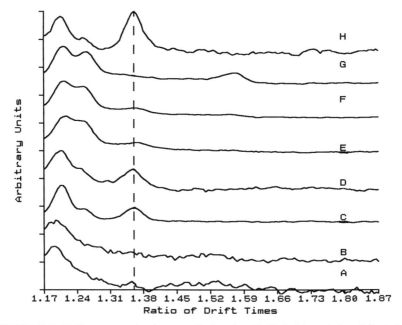

FIGURE 10.5. IMS spectra of mixtures of phenol and bis-2-chloroethyl sulfide before application of an FIR digital filter.

in a stacked spectrum plot in Fig. 10.5. Each spectrum in the plot has been scaled such that the maximum peak height of any one spectrum is the same as the maximum peak height of all the other spectra. These peak intensities have not been changed, only the plot size has been changed. The dashed line in Figs. 10.5 and 10.6 is located at a T_R value of 1.36. This is the value at which one would expect to find the bis-2-chloroethyl sulfide ion peak.

An IFFT has been applied to each spectrum which is contained in the database to create an interferogram. A digital filter, corresponding to a Gaussian-shaped peak which is four data points wide and centered at 1.36 on the T_R axis, was applied to the interferogram. After the digital filter, an FFT was performed to give a modified IMS spectrum. A stacked plot of all eight of the spectra, after application of the digital filter, is shown in Fig. 10.6. Again, each spectrum in the plot has been graphically scaled such that the height of the most intense peak in any one spectrum is the height of the most intense peak in all of the other spectra in the plot.

Comparison of Figs. 10.5 and 10.6 show the enhancement that has been gained as a result of the digital filter. In Fig. 10.6, spectra C, D, and H have, as the major spectral feature, a peak at $T_R = 1.36$. The corresponding spectra in Fig. 10.5 exhibit only minor spectral features at this value of the T_R. Spectra E and F in Fig. 10.6 have a peak at a T_R of 1.36

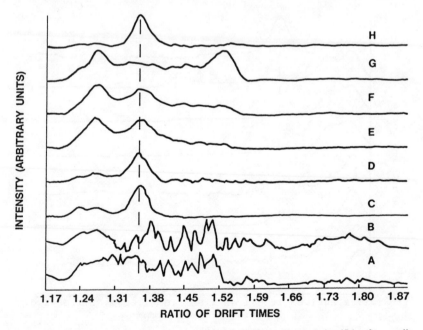

FIGURE 10.6. IMS spectra of mixtures of phenol and bis-2-chloroethyl sulfide after application of an FIR digital filter.

as one of the two major spectral features, while the corresponding spectra in Fig. 10.5 exhibit only minor spectral features. Spectrum G in Fig. 10.6 exhibits what may be a peak at the T_R value of 1.36; however, identification of this feature as the bis-2-chloroethyl sulfide peak at this time would be difficult. Spectrum G in Fig. 10.5 shows no spectral feature at the proper location; thus, identifying the presence of bis-2-chloroethyl sulfide is impossible. Finally, spectra A and B shown in Figs. 10.5 and 10.6 have only a high noise component after application of the digital filter, as shown in Fig. 10.6. This is because no peak is present in the original spectra within the bandpass of the digital filter.

10.4. LINEAR DISCRIMINANT ANALYSIS

In recent years, the use of IMS for the analysis of airborne vapors has been increasing.[3-7] However, much of the effort associated with the analysis of the IMS spectra has been left to the chemist. In an effort to aid in the preliminary identification, a personal computer (PC) based spectrum identification package has been developed. This package, written in Microsoft Fortran, uses a linear discriminant function for its identification, and consists of three separate programs. These programs are: IMSDISC, a program which reads selected data files from the PC and builds a discrimination data set; TRAIN, a program which analyzes the discrimination set and calculates the linear discriminant function that best isolates the data of interest from the interferent data; and IMSIDENT, a program which reads the data to be analyzed and identified and calculates its linear discriminant value.

Linear discriminant analysis, one of the most basic forms of pattern recognition used by scientists, is used as a supervised learning technique. In supervised learning techniques, the computer learns to classify the samples being analyzed based on knowledge about the samples; in this study, the samples either belong to the class of chemicals you wish to identify, or they do not. The goal of the learning is to develop a classification rule, the linear discriminant function, which allows the validity of the classification to be tested and ultimately to properly classify unknowns.

The linear discriminant function has the general form

$$g(x) = w_0 + \sum_{i=1}^{n} w_i x_i \qquad (10.6)$$

where w_0 is the threshold vector, w_i the weight vector, x_i the response vector, and $g(x)$ is the response function. The discriminant function $g(x)$ is determined by choosing those variables x_i with characteristics which differ between the groups being classified. These variables are then combined linearly and weighted such that the groups are as statistically different as possible. This linear combination of variables is calculated using the perceptron convergence criteria.

The perceptron[11-13] is a pattern recognition procedure which consists of updating the weight vector by considering only those patterns, or spectra in this work, that have been misclassified in the training set. Each misclassified pattern is considered in turn, with a fraction of each misclassified spectrum being added to the weight vector. This procedure is continued until all of the spectra are classified correctly, or until it is determined that the procedure fails to converge to a satisfactory solution.

In this software package, the three programs are run separately, but are still interrelated. The first program, IMSDISC, uses a file called NAMES, which is simply the file that contains the names of the individual data files to read, and a value that tells the program whether the file is to be treated as the sample or as an interferent. The data from the individual data files is then treated such that all the files are compatible with respect to time spacing between data points, delay to start of data sampling, and number of data points. To accomplish this, IMSDISC uses a spline function to interpolate and fit the data. After the data has been treated to fill the compatibility requirement, the discriminant threshold is set to zero by multiplying all interferent spectra by negative 1 (-1). The sample spectra are left unaltered. The data is then stored in a discriminant data file.

The second program, TRAIN, develops a linear discriminant based on the perceptron convergence criteria. TRAIN prompts the operator for the name of the input discriminant file that was created with the program IMSDISC. It reads the data from the discriminant data set, accepts input for the values of a scaling factor, between 0.000000001 and 0.1, and the number of iterations to perform using this scaling factor. In practice, it is generally necessary to use a series of decreasing scaling factors and iterations to calculate the linear discriminant function which best differentiates the samples and the interferents. After the linear discriminant function has been calculated, the linear coefficients are written to a file on the computer disk for use by the last program. These first two programs, IMSDISC and TRAIN, are the time-consuming programs and are run only when a new compound is to be added to the database.

The third program in this package, IMSIDENT, uses the linear discriminant values created with the program TRAIN. Thus, it is dependent on the first two programs in the package. IMSIDENT can be used in one of two possible configurations: the first configuration is as a stand-alone program, and the second is that it can be incorporated into a data collection program for real-time identification of an unknown environment. In the stand-alone configuration, the program prompts the operator for the name of the data to analyze. The program reads the data, and performs a spline interpolation to make the data compatible with the discriminant data sets. Next, the program reads a file named COEF.FIL that contains the names of the coefficient files. The linear discriminant value is then calculated. If the linear discriminant value is positive, an alarm message is generated which notifies the operator that the spectrum has been identified. No message is generated if the discriminant value is negative. The results of the identification process are then written to a file named ALARM.RPT for later use, and the program then prepares to read the next data file to be analyzed.

In the second configuration, the program functions as a real-time monitor. The name of the data file to be analyzed is passed from the data collection program to the IMSIDENT package, rather than prompting the operator for the name of the data file to analyze. The spline interpolation is then performed on the data, and the linear discriminant value is calculated. If the discriminant value is positive, the alarm message is generated; no message is generated if the discriminant value is negative. The results of the identification process are written to a file named ALARM.RPT for later use.

Discussion

The program package was developed for use with the Graseby Ionics Advanced Signal Processing (ASP) Board, the Graseby Airborne Vapor Monitor (AVM), and a Zenith 286 PC. Using this hardware and the linear discrimination package, it has been possible to identify and semiquantitate the presence of 11 common chemical vapors in air. These compounds, most of which are of industrial importance, are diethyl ether, acetone, phenyl-2-propanone, methanol, isooctane, cyclohexanone, acetic anhydride, piperidine, hydrogen iodide, methyl ethyl ketone, and methyl isobutyl ketone. These compounds, and the levels at which the Occupational Safety and Health Administration (OSHA) have determined them to be hazardous, are shown in Table 10.2. When the software is used in the stand-alone configuration (i.e., separate from the data collection routines) and using the Zenith 286 PC, the presence of these compounds can be determined and the compound identified in less than ten seconds. This includes the time necessary to perform the spline interpolation and the calculation of the discriminant value for the data; however, this does not include the time required to create the discriminant functions.

The results shown in Table 10.3 are from the evaluation of a series of files used to determine the presence of N-methyl formamide. The "All Clear" report indicates that the IMSIDENT program does not find any similarities between the N-methyl formamide test spectrum and the spectra

TABLE 10.2. Compounds Being Analyzed Using
Linear Discriminant Analysis and Their Action Levels
in Parts per Million[14–16] a

Compound	TWA	STEL	Ceiling	IDLH
Acetic acid	10	15	NA	1000
Acetic anhydride	5	NA	5	1000
Acetone	750	1000	NA	20000
Benzene	1	NA	1	NA
Cyclohexanone	25	100	NA	5000
Diethyl ether	400	500	NA	19000
Formamide	20	30	NA	NA
Hydroiodic acid	NA	NA	NA	NA
Iodine	NA	0.1	0.1	10
Isooctane	300	375	NA	5000
Isopropanol	400	500	NA	12000
Methanol	200	250	NA	25000
Methyl ethyl ketone	200	300	NA	3000
Phenyl-2-propanone	NA	NA	NA	NA
Toluene	100	150	NA	2000

a The Time Weighted Average (TWA), the Short-Term Exposure Limit (STEL), the
Ceiling Limit (Ceiling), and the Immediate Danger to Life and Health (IDLH) are
defined in Ref. 14. A value reported as NA indicates no action level.

TABLE 10.3. File "ALARM.RPT" for N-Methyl Formamide Analysis

```
DISCRIMINANT VALUE =        576.796000
ETHER   ******* ALARM ******  ON FILE       \AVM\DATA\nmfo0000.ACQ
DISCRIMINANT VALUE =        548.082000
ETHER   ******* ALARM ******  ON FILE       \AVM\DATA\nmfo0001.ACQ
DISCRIMINANT VALUE =        580.730500
ETHER   ******* ALARM ******  ON FILE       \AVM\DATA\nmfo0002.ACQ
DISCRIMINANT VALUE =        510.498600
ETHER   ******* ALARM ******  ON FILE       \AVM\DATA\nmfo0003.ACQ
DISCRIMINANT VALUE =        660.293000
ETHER   ******* ALARM ******  ON FILE       \AVM\DATA\nmfo0004.ACQ
      ALL CLEAR FOR FILE      \AVM\DATA\nmfo0005.ACQ
      ALL CLEAR FOR FILE      \AVM\DATA\nmfo0006.ACQ
      ALL CLEAR FOR FILE      \AVM\DATA\nmfo0007.ACQ
      ALL CLEAR FOR FILE      \AVM\DATA\nmfo0008.ACQ
      ALL CLEAR FOR FILE      \AVM\DATA\nmfo0009.ACQ
      ALL CLEAR FOR FILE      \AVM\DATA\nmfo0010.ACQ
      ALL CLEAR FOR FILE      \AVM\DATA\nmfo0011.ACQ
      ALL CLEAR FOR FILE      \AVM\DATA\nmfo0012.ACQ
      ALL CLEAR FOR FILE      \AVM\DATA\nmfo0013.ACQ
      ALL CLEAR FOR FILE      \AVM\DATA\nmfo0014.ACQ
      ALL CLEAR FOR FILE      \AVM\DATA\nmfo0015.ACQ
```

of the eleven compounds stored in the database. The report of an alarm indicates that the program did find similarities in the spectra, and the magnitude of the discriminant is a measure of the amount of similarity.

It is not really surprising that there are a number of false positive alarms indicating the presence of diethyl ether. Older versions of the AVM used an acetone dopant within its detection system, while newer versions of the AVM use water vapor in the atmosphere as the dopant. This dopant in the older AVMs results in the presence of an acetone reactant ion. This reactant ion is the ionic species responsible for transferring the ionic charge to the chemical compound being studied. All of the spectra used in the discrimination functions were recorded using water as the reactant ion. Thus, the discriminant functions have not been trained to eliminate the possibility of alarming on a spectrum which has an acetone reactant ion peak, and an alarm is reported. Examination of two representative spectra for which an alarm was reported shows the similarity of the IMS spectrum for the diethyl ether, the lower trace in Fig. 10.7 (ETHER in Table 10.3) and N-methyl formamide background spectrum, the upper trace in Fig. 10.7 (\AVM\DATA\nmfo0000.ACQ in Table 10.3). The location of the reactant ion peak does not appear at the same time as does the diethyl ether peak, however the band shapes are similar. If the discriminant function is trained to ignore the acetone reactant ion peak, one does not get an alarm. Results of the identification procedure with the acetone reactant ion peak being ignored is shown in Table 10.4.

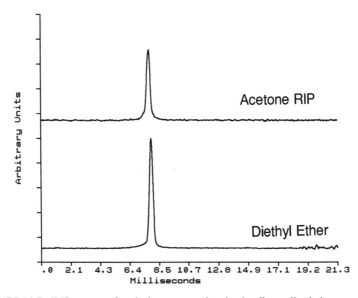

FIGURE 10.7. IMS spectra of typical spectra analyzed using linear discriminant analysis.

TABLE 10.4. File "ALARM.RPT" for
N-Methyl Formamide Analysis with
Acetone Reactant Ion Ignored

ALL CLEAR FOR FILE	\AVM\DATA\nmfo0000.ACQ
ALL CLEAR FOR FILE	\AVM\DATA\nmfo0001.ACQ
ALL CLEAR FOR FILE	\AVM\DATA\nmfo0002.ACQ
ALL CLEAR FOR FILE	\AVM\DATA\nmfo0003.ACQ
ALL CLEAR FOR FILE	\AVM\DATA\nmfo0004.ACQ
ALL CLEAR FOR FILE	\AVM\DATA\nmfo0005.ACQ
ALL CLEAR FOR FILE	\AVM\DATA\nmfo0006.ACQ
ALL CLEAR FOR FILE	\AVM\DATA\nmfo0007.ACQ
ALL CLEAR FOR FILE	\AVM\DATA\nmfo0008.ACQ
ALL CLEAR FOR FILE	\AVM\DATA\nmfo0009.ACQ
ALL CLEAR FOR FILE	\AVM\DATA\nmfo0010.ACQ
ALL CLEAR FOR FILE	\AVM\DATA\nmfo0011.ACQ
ALL CLEAR FOR FILE	\AVM\DATA\nmfo0012.ACQ
ALL CLEAR FOR FILE	\AVM\DATA\nmfo0013.ACQ
ALL CLEAR FOR FILE	\AVM\DATA\nmfo0014.ACQ
ALL CLEAR FOR FILE	\AVM\DATA\nmfo0015.ACQ

10.5. CONCLUSIONS

A significant improvement in the ability to recognize spectral features has been shown as a result of applying a simple FIR digital filter to IMS data. It has been demonstrated that for cases in which materials of interest (such as bis-2-chloroethyl sulfide and phenol) are mixed and the ratio of the bis-2-chloroethyl sulfide concentration to phenol concentration ranges from 2:1 to 1:2.5, one can resolve the presence of and identify the bis-2-chloroethyl sulfide peak. This is an improvement over the current baseline correction/peak detection algorithm used in the atmospheric vapor monitor. In more extreme cases, where the ratio of the bis-2-chloroethyl sulfide concentration to the phenol concentration is on the order of 1:5, less noticeable improvement can be seen. For moderate concentrations with mixture ratios less than 1:5, it appears that the bis-2-chloroethyl sulfide signal does not exist in the data. It might well be that no amount of signal processing will extract the bis-2-chloroethyl sulfide features when phenol concentrations are greater than 5 times the bis-2-chloroethyl sulfide concentrations. This preliminary study indicates a potential for using digital filters to enhance the identification of IMS spectral features represented by small, narrow features adjacent to large, broad interferent features. Future research conducted by this laboratory in this area will be directed to the application of more sophisticated digital features than those used in this study.

The use of linear discriminant functions to analyze ion mobility spectra has also been shown to be a useful tool for spectral identification. The data analysis package is versatile, yet powerful. Simple modifications to the spectral library can result in the rejection of unwanted alarms; yet other spectra which exhibit characteristic bands similar to those in the alarm data sets can still be identified. The portions of the program that deal with data retrieval can be modified to accept input which differs from that used by the Graseby ASP Board, without altering the rest of the program. This enables the software package to be used on a wide variety of data. The Fortran listings of the programs are available from the address listed above.

ACKNOWLEDGMENTS

The authors wish to thank Dr. Gary Small and co-workers at the University of Iowa for providing one of the software packages (the software package IFILES) used in this study. This package was modified for use with the ion mobility data. In addition, the authors wish to thank Mr. John Ditillo of the Detection Directorate, U.S. Army Chemical Research, Development, and Engineering Center for providing several subroutines used in the software package IMSD, Mr. Donald B. Shoff for providing the ion mobility spectra for evaluating the software package, and Dr. Charles Harden and Mr. David Blyth for their helpful comments and discussions during the course of this work.

REFERENCES

1. E. W. McDaniel and E. A. Mason, *The Mobility and Diffusion of Ions in Gases*, Chapter 2, Wiley, New York (1973).
2. H. E. Revercomb and E. A. Mason, *Anal. Chem.* **47**, 970 (1975).
3. T. W. Carr, ed., *Plasma Chromatography*, Plenum Press, New York (1984).
4. G. A. Eiceman, A. P. Snyder, and D. A. Blyth, *Int. J. Environ. Anal. Chem.* **38**, 415 (1989).
5. G. A. Eiceman, M. E. Fleischer, and C. S. Leasure, *Int. J. Environ. Anal. Chem.* **28**, 279 (1987).
6. C. S. Leasure, M. E. Fleischer, and G. A. Eiceman, *Anal. Chem.* **58**, 2141 (1986).
7. J. M. Preston and L. Rajadhyax, *Anal. Chem.* **60**, 31 (1988).
8. D. Childers and A. Durling, *Digital Filtering and Signal Processing*, West Publ., St. Paul, MN (1975).
9. G. W. Small, R. T. Kroutil, J. T. Ditillo, and W. R. Loerop, *Anal. Chem.* **60**, 264–269 (1988).
10. R. T. Kroutil, J. T. Ditillo, and G. W. Small, in: *Computer-Enhanced Analytical Spectroscopy*, Vol. 2 (H. Meuzelaar, ed.), pp. 71–111, Plenum Press, New York (1989).
11. F. Rosenblatt, *Principles in Neurodynamics: Perceptrons and the Theory of Brain Mechanisms*, Spartan, New York (1962).

12. R. O. Duda and P. E. Hart, *Pattern Classification and Scene Analysis*, Wiley, New York (1973).
13. Y.-H. Pao, *Adaptive Pattern Recognition and Neural Networks*, Addison-Wesley, New York (1989).
14. *Threshold Limit Values and Biological Exposure Indices for 1989–90*, American Conference of Governmental Industrial Hygienists, Cincinnati, OH (1989).
15. Code of Federal Regulations, 29 CFR 1910, Subpart Z—Toxic and Hazardous Substances, 1910.1000—Air Contaminants (19 January 1989).
16. *NIOSH Pocket Guide to Chemical Hazards*, U.S. Department of Health and Human Resources, National Institute of Occupational Safety and Health, Washington, DC (1985).

Computerized Multichannel Atomic Emission Spectroscopy

11

Robert B. Bilhorn, R. S. Pomeroy, and M. B. Denton

11.1. ATOMIC EMISSION SPECTROSCOPY—COMPLICATIONS BROUGHT ABOUT BY COMPLEX SAMPLES

Elemental analysis by atomic emission spectroscopy is widely applicable and useful for determinations ranging from part-per-billion levels to percent levels. Several source options are available making the technique amenable to solids, liquids, and gases. The subject of analytical emission spectroscopy is discussed in detail in several excellent reference texts. The interested reader is referred to Boumans's two-book series on inductively coupled plasma emission spectroscopy[1,2] which also provides a detailed guide to the literature on emission spectroscopic analysis using other sources. A very brief overview of emission spectroscopy will be presented here for the purpose of providing a foundation for the following discussion.

Analytical atomic emission sources produce complex spectra which carry a wealth of information about the sample and the source. The usable emission from plasmas, arcs, and sparks extends from the vacuum

Robert B. Bilhorn • Analytical Technology Division, Eastman Kodak Company, Rochester, New York 14652-3708.   R. S. Pomeroy and M. B. Denton • Department of Chemistry, University of Arizona, Tucson, Arizona 85721.

Computer-Enhanced Analytical Spectroscopy, Volume 3, edited by Peter C. Jurs. Plenum Press, New York, 1992.

ultraviolet to the near-infrared spectral regions; however, only a portion of the vacuum ultraviolet, the ultraviolet, and visible regions is commonly used. Spectral linewidths vary from 1 to 10 pm (picometer) and a single element may emit thousands of spectral lines. Quantitative analysis is carried out using either internal or external standards and the dynamic range of signal intensities can cover six orders of magnitude.

In principle, a single spectral line is all that is necessary to identify the presence of an element and determine its concentration. In practice this is often the case and many analytical procedures have been developed and successfully applied using a single spectral line per element. As the matrix composition becomes more complex and the analysis requirements become more demanding, however, more of the information available in the emission spectrum must be used. For example, the working curve for the most sensitive spectral line of an element may become nonlinear at high concentrations. For precise work at high concentrations a less sensitive spectral line which produces a linear working curve may be more appropriate.

In complex matrices the possibility of encountering spectral inter-ferences increases.[3,4] The direct overlap of a spectral line by a line from a concomitant species requires either the use of a different analytical line or correction for the contribution to the overall intensity by the concomitant element. Other types of interferences on the intensities of analyte spectral lines also occur. The emission from molecular species present in the source takes the form of a more broad spectral background which can contain structure and varies depending on the sample composition. The proximity of an analyte line to an intense matrix or source line, such that the wing of the intense line produces an interference, can produce a variable and nonlinear baseline. Instrumental considerations can also play a role. For example, scattered and stray light contributions to the spectral background depend on the intensity and complexity of the sample matrix emission. In all of these cases it is desirable to measure not only the intensity of the analyte spectral line but also the intensity of the spectral background on both sides of the spectral line. Time-varying background can only be corrected for if the background and analyte emission-line-intensity measurements are conducted simultaneously. When direct spectral inter-ferences occur, the flexibility to use more than one spectral line for an element is also desirable.

Modern plasma sources, such as the inductively coupled plasma (ICP) and the direct current plasma (DCP), have set new standards for accuracy and precision in widely varying matrix types. Although touted as being essentially free from matrix effects, variations between standards and samples can produce variations in the amount of analyte entering the plasma as well as changes in the excitation conditions which exist in the plasma.[5-8] Changing sample introduction rates and plasma excitation conditions lead to variations in emission intensities that are independent of the concentration of the analyte in the original sample. Here, even more information from the emission spectrum must be used to achieve correct results. Recent work has demonstrated that emission features due to the plasma gas itself, and to the solvent being introduced with the sample, provide indicators as to the prevailing sample introduction and excitation conditions.[8] The work suggests that these spectral features may be used as feedback signals for altering plasma conditions so as to restore the desired correspondence between emission intensity and analyte concentration.

It is clear from these considerations that the flexibility to use a large part of the information available in the atomic emission spectrum is important when the best possible accuracy and precision are required.

11.2. ATOMIC EMISSION INSTRUMENTATION

Analytical emission spectroscopy was first performed using photographic emulsions as the detection mechanism. Plates or strips of film provided continuous coverage of the spectrum, and multiple exposures, made at differing integration times, were used to cover the dynamic range of the source. Direct electronic readout, made possible by the advent of phototubes, and then photomultiplier tubes (PMTs), has essentially replaced photographic readout. The disadvantage of no longer providing coverage of the complete emission spectrum in a single exposure is outweighed by the advantages of an increased linear dynamic range and immediate availability of results. Two general approaches are used to provide the multiwavelength information required in emission spectroscopy with PMTs: a sequential approach, which provides the wavelength information encoded in time, and a simultaneous approach, which provides the wavelength information encoded in space.

11.2.1 Slew-Scanning Monochromators

Slew-scanning monochromators use a moving grating to bring individual spectral lines to focus on a single photomultiplier tube (PMT) detector. These instruments allow access to any wavelength in the spectrum commonly used for atomic emission analysis but, by their nature, only allow the observation of a single spectral line at a time. Slew-scanning monochromators are successful with temporally stable sources such as the ICP and DCP but obviously are unsuitable for arc and spark discharges where a second, internal standard emission wavelength must be monitored simultaneously for quantitative work. Slew-scanning monochromators require additional time as the number of elements to be determined by plasma emission increases, requiring greater temporal stability from the source. In sample limited situations, a slew-scanning monochromator may not be fast enough when a large number of elements must be determined.[9] Slew-scanning monochromators offer a considerable advantage however—the flexibility to use the best spectral line for an analysis depending on the matrix composition and the analyte concentration.

11.2.2. Polychromators

Simultaneous multielement analysis by plasma, arc, and spark emission is carried out using polychromators. The placement of upward of 50 slits and PMTs around the focal curve of a fixed grating spectrograph allows the simultaneous determination of a large fraction of the elements. The laborious process of aligning the slits in the focal plane prevents flexibility in changing spectral lines and, although many elements may be determined simultaneously, only a very small fraction of the total information available in the spectrum is ever used. Unlike slew-scan monochromators, the fixed nature of the exit slits sometimes results in a less-than-optimum spectral line being used for an analysis. Modern polychromators do have the capability to scan the spectral background in the vicinity of the analytical lines, however, so interference detection and spectral background correction are possible.

Several variations on the two basic experimental approaches have been developed to combine some of the advantages of each. These instruments do not offer true simultaneous coverage of the entire emission spectrum but offer increased flexibility and speed of analysis. Some of the hybrids include slew-scan monochromators added as a channel to a polychromator, instruments with interchangeable cassettes of slits and PMTs at the focal plane of a polychromator,[10] instruments with multiple moving detectors at the focal plane of a polychromator, scanning polychromators,[11] and polychromators which put only specially selected regions of the spectrum on a linear multichannel detector at the focal plane.[12,13] The increased performance of these systems is of course accompanied by added complexity. As more demands are placed on the performance of atomic emission instrumentation, the trend seems to be one of evolution back toward continuous coverage of the analytically useful spectrum with the ultimate solution being the electronic equivalent of a photographic emulsion.

11.3. MULTICHANNEL DETECTORS FOR ATOMIC EMISSION SPECTROSCOPY

The advantages of having continuous coverage of the entire emission spectrum have been recognized for years and have led to the study of two-dimensional multichannel arrays originally designed for television applications.[14] The requirements for operation in the ultraviolet, high sensitivity, and very large dynamic range have prevented a workable solution from being arrived at until the advent of sensors based on silicon integrated-circuit technology. Even so, only special variations of these have been able to meet all of the requirements necessary to make multichannel arrays a viable alternative to PMTs.

The class of image detectors called charge-transfer devices (CTDs) have so far demonstrated the most promise for atomic emission spectroscopy. The two members of the class, so named because photogenerated charge is moved or transferred in the silicon for measurement, are charge injection devices (CIDs) and charge-coupled devices (CCDs).[15–18] Although closely related, the two devices differ in the method by which the photogenerated charge is measured, and it is this difference which allowed the CID to first meet the dynamic-range requirement which proved to be the most significant impediment to the wide application of multichannel detectors.

11.3.1. The Charge Injection Device

11.3.1.1. CID Architecture and Photogenerated Charge Readout

Both CIDs and CCDs store photogenerated charge in individual elements of the detector called pixels. Potential wells for charge storage are produced in the silicon by the application of voltages to doped polycrystalline electrodes which overlie the imaging area. At the termination of an exposure to a scene, the quantity of charge in each pixel is measured. In CIDs this can be done nondestructively, that is, the quantity of charge can be determined without removing any of it from the pixel. The procedure involves shifting the photogenerated charge back and forth between two potential wells produced by two crossed electrodes (see Fig. 11.1). During charge integration, the column electrode is biased negatively so that photogenerated charge (holes in the CID) is collected under the column electrode. For readout, the row electrode is set to an intermediate voltage through momentarily closing the row reset switch (Fig. 11.1b). The resulting voltage is sampled by the amplifier, which is connected to a single row through the row select scanner. A positive voltage pulse applied to a single column, connected through the action of the column select scanner, causes the photogenerated charge along that column to be shifted to the corresponding row potential wells. A voltage change is induced along all of the rows and is measured for the row connected to the amplifier. In this way, the photogenerated charge stored at the intersection of the selected row and

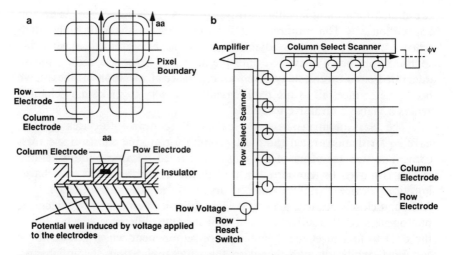

FIGURE 11.1. Construction of a charge injection device. (a) Plan and cross-section views of a pixel showing the crossed row and column electrodes which overlie the charge accumulation region. (b) Schematic showing the row and column scanners used to access individual pixels. Also shown is the switch used to establish the row voltage.

column is measured. A voltage signal proportional to the amound of stored charge is obtained by subtracting the voltage measured prior to charge transfer.

The readout procedure can be repeated practically indefinitely in the CID simply by restoring the negative column voltage, thus allowing the photogenerated charge to return to its original position. This is followed by reestablishing the row voltage by momentarily closing the row reset switch, and repeating the readout procedure described above.

The horizontal and vertical scanners in a CID can be run independently, and rows and columns can be skipped over. This feature allows small sections of the array to be accessed without the need to readout the entire array. The scanners access rows and columns sequentially rather than truly randomly, however, so the capability is called pseudorandom access.

11.3.1.2. Dynamic-Range Extension for Use in AES

Nondestructive readout (NDRO) is used with the CID to reduce random noise, making small quantities of charge detectable, and to extend the dynamic range of the sensor. The difference between the smallest level of charge measurable and the largest amount of charge that can be stored in a CID pixel is not sufficiently large to match the dynamic range of AES sources. Fortunately, when excess charge is added to a CID pixel, it does not spill into adjacent pixels (bloom) as is common in many detector arrays. Therefore, different exposure times may be used to allow measurement of both very intense and very faint spectral lines. The beauty of the NDRO capability is that the exposure time for every spectral line can be different. A procedure called random-access integration (RAI) has been developed[21] in which NDROs are used during a single exposure to follow the accumulation of charge at both weak and intense spectral lines. NDROs are used continuously during the exposure to determine when a high signal-to-noise ratio measurement is possible at each spectral line. Once a sufficient amount of charge has accumulated at a spectral line position, its exact quantity is measured and the exact exposure time is recorded. The dynamic range of the CID combines with the variable integration time to produce the dynamic range needed for AES. The RAI procedure is the most efficient means of measuring the intensity of a number of spectral lines, because the exposure time is adjusted dynamically according to signal-to-noise requirements based on the intensity of each spectral line.

11.3.2. The Charge-Coupled Device

11.3.2.1. CCD Architecture and Photogenerated Charge Readout

The readout process used in CCDs involves the transfer of photogenerated charge (electrons) to a specialized amplifier located on the periphery of the integrated circuit. At the termination of an exposure, charge in the entire array is shifted in unison along the direction of the columns. This is accomplished by progressively advancing a potential barrier by the application of voltages to the overlying electrodes as shown in Fig. 11.2 for a hypothetical three-phase CCD. The boundary between columns is a permanent barrier called a channel stop and the boundary between rows of pixels is a movable barrier produced by the voltages applied to the electrodes. When the charge in the array is shifted, the row at the edge of the array is shifted into a transport register (serial register) designed to allow the transfer of charge in that single row toward the output amplifier. The amplifier is located at the end of the serial register and allows the sequential measurement of the quantity of charge in each pixel along the row. The photogenerated charge in the entire array is readout by shifting one row at a time into the serial register and then shifting each row to the amplifier.

Charge-coupled devices achieve lower noise levels than CIDs, primarily because the charge is transferred to the amplifier rather than transferred to a long row electrode which is in turn connected to an amplifier. The amount of capacitance at the input of the amplifier determines to a large

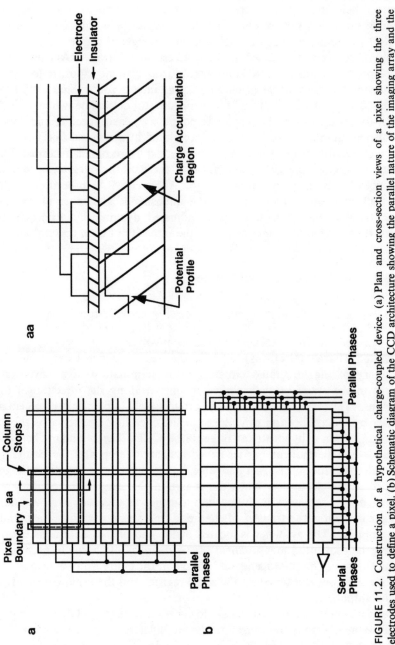

FIGURE 11.2. Construction of a hypothetical charge-coupled device. (a) Plan and cross-section views of a pixel showing the three electrodes used to define a pixel. (b) Schematic diagram of the CCD architecture showing the parallel nature of the imaging array and the serial register and output amplifier.

extent the noise. In spite of the lower noise levels, CCDs like CIDs do not have a sufficiently large dynamic range for AES. More importantly, most CCDs bloom. That is, when the dynamic range of a pixel is exceeded, the excess charge spills into adjacent pixels. Blooming usually occurs preferentially along columns until charge is spilled into the serial register. Charge then spills along the serial register and back into other columns until the entire array is saturated with charge. Several schemes have recently been devised to control blooming to some extent in conventional CCDs[22-25]; however, their ability to cope with severe optical overloads has not yet been established.

11.3.2.2. Antiblooming CCDs for Wide-Dynamic-Range Detection in AES

Several CCDs have been produced over the years which do not bloom. Structures are integrated into each pixel and are designed to conduct away excess charge and prevent it from spilling into adjacent pixels. An antiblooming CCD is suitable for use in AES, because long exposures can be used for the measurement of weak spectral lines without fear of blooming occurring from intense spectral lines. Because no NDRO capability exists for CCDs, separate exposures must be used for intense and faint spectral lines; but several advantages over CIDs also exist. The primary advantage is that antiblooming CCDs are currently commercially available in camera systems suitable for AES. By contrast, a CID camera system which allows cooling (the need for which is described in the following section), NDROs, and subarray readout is not commercially available and must be built in the laboratory. The commercially available antiblooming CCDs also offer more pixels, a larger dynamic range within a single exposure (intrascenic dynamic range), and better low-light-level sensitivity.

11.4. EXPERIMENTAL CID- AND CCD-BASED POLYCHROMATORS

11.4.1. Echelle Spectrometers for Two-Dimensional Detector Arrays

The two-dimensional format of CIDs and CCDs, although providing for a large number of pixels in a relatively small area of silicon, requires a compatible polychromator focal plane format. Polychromators based on an echelle grating can produce a two-dimensional format and offer the added advantage of providing high resolution in a relatively compact instrument.[26,27] Echelle gratings are coarsely ruled, have steep blaze angles, and operate in multiple high orders. As a result, they produce high

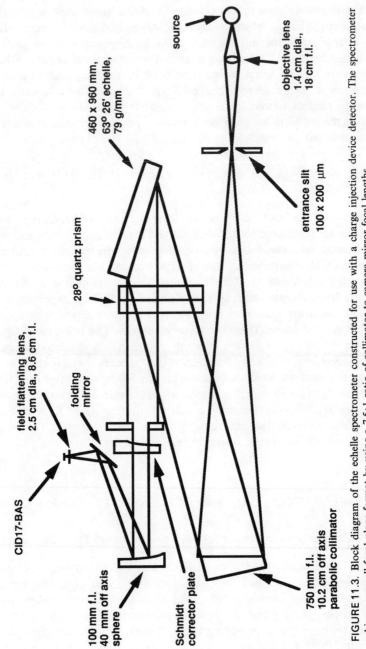

source

objective lens
1.4 cm dia.,
18 cm f.l.

entrance slit
100 × 200 μm

460 × 960 mm,
63° 26' echelle,
79 g/mm

28° quartz prism

field flattening lens,
2.5 cm dia., 8.6 cm f.l.

folding
mirror

CID17-BAS

100 mm f.l.
40 mm off axis
sphere

Schmidt
corrector plate

750 mm f.l.
10.2 cm off axis
parabolic collimator

FIGURE 11.3. Block diagram of the echelle spectrometer constructed for use with a charge injection device detector. The spectrometer achieves a small focal plane format by using a 7.5:1 ratio of collimator to camera mirror focal lengths.

angular dispersion but have a very limited free spectral range. Echelle gratings are usually used with a prism or a second grating operated in first order to separate the overlapping echelle orders. The direction of the order sorter dispersion is set normal to the direction of the echelle dispersion producing a roughly rectangular focal plane which appears as a series of linear spectra arranged in rows like lines of text on a page. Each row covers a different piece of the spectrum, and the wavelength varies continuously as you move along in a row as well as down through the rows. Two complete spectrometer systems, one based on a CID detector and one based on an antiblooming CCD detector, have been developed. The two instruments take two different approaches to produce a focal plan format compatible with the respective detectors. Each offers advantages and disadvantages.

11.4.1.1. A Focal Plane Image Demagnifying Spectrometer

The CID-detector-based system has been described in detail else-where[21] and consists of an echelle spectrometer with a quartz prism for cross dispersion. An image size that matches the size of the CID is achieved from a relatively high angular dispersion grating and prism combination by using a camera mirror focal length which is 7.5 times shorter than the collimating mirror focal length. Details of the optical design are shown in Fig. 11.3. The spectrometer covers a continuous wavelength range extending from approximately 220 to 515 nm. The aluminum 309 nm doublet (309.2713 nm and 309.2842 nm) is partially resolved by this instrument as shown in Fig. 11.4.

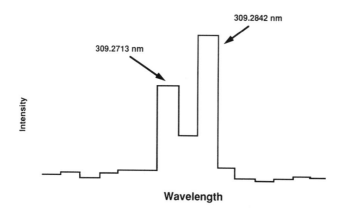

FIGURE 11.4. Intensity profile across the aluminum 309.3 nm doublet recorded with the CID17-BAS echelle spectrometer using a direct-current argon plasma source. The 0.013 nm separated lines are partially resolved.

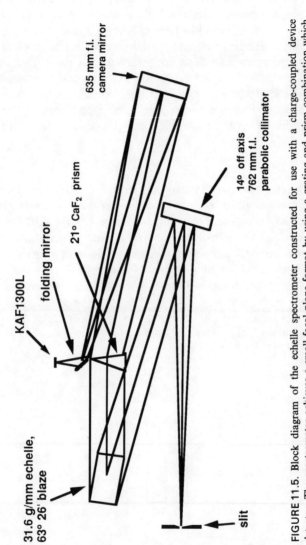

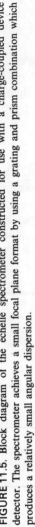

635 mm f.l.
camera mirror

21° CaF$_2$ prism

KAF1300L
folding mirror

14° off axis
762 mm f.l.
parabolic collimator

31.6 g/mm echelle,
63° 26' blaze

slit

FIGURE 11.5. Block diagram of the echelle spectrometer constructed for use with a charge-coupled device detector. The spectrometer achieves a small focal plane format by using a grating and prism combination which produces a relatively small angular dispersion.

11.4.1.2. A Low-Angular-Dispersion High-Resolution Spectrometer

The instrument based on the CCD detector also uses an echelle grating with a prism for cross dispersion. An image size which is compatible with the size of the CCD is achieved by using a more coarsely ruled echelle grating. The more coarse ruling produces more echelle orders and a lower angular dispersion within an order, so that a longer focal length camera mirror can be used to produce the same size image. Figure 11.5 shows the details of the optical system, which includes an off-axis parabolic collimator, a calcium fluoride prism, and a spherical camera mirror. The echelle grating is illuminated in the plane which is perpendicular to the plane of diffraction in order to minimize the amount of light lost due to shadowing by the deep echelle grooves.[28] The spectrometer covers a wavelength range extending from approximately 180 to 465 nm and achieves baseline resolution of the mercury 313 nm doublet as shown in Fig. 11.6.

11.4.2. Slow-Scan Charge-Transfer Device Cameras

The cameras built for operating CIDs and CCDs in spectroscopic and scientific imaging applications are still cameras, as opposed to video cameras, which readout the array at fixed (relatively short) time intervals. A mechanical shutter is used to control the continuously variable exposure

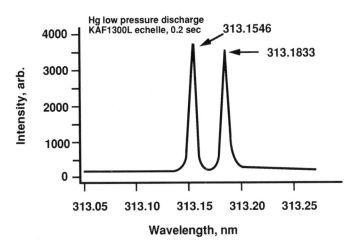

FIGURE 11.6. Intensity profile across the mercury 313 nm doublet recorded with the KAF1300L echelle spectrometer using a low-pressure discharge lamp source. The 0.0287 nm separated lines are baseline resolved.

time and several-minute exposures are routinely used to achieve very good low-light-level sensitivity. Charge which is thermally generated in the silicon becomes significant at long exposure times, so some form of cooling is typically provided. If the arrays were not cooled, some of the dynamic range would be consumed by the dark current but, more importantly, signal-to-noise would be degraded by the dark-current shot noise. The pixel intensity is digitized directly to 12, 14, or, in some cases, 16 bits of precision. To achieve very low noise levels, the rate of pixel readout is much slower than in a video camera, hence the name slow-scan. Pixel rates ranging from 50 to 500 kHz are typical, so the time required to read an entire array can vary from one to 20 seconds depending on the size of the array and the readout speed.

11.4.2.1. CID17-BAS Camera System

The charge injection device camera used in these studies is based on the CID17-BAS sensor (CID Technologies, Liverpool, NY) and was constructed in-house.[29] The camera is operated by a Photometrics Ltd. camera controller (Model CC183, Tucson, AZ) with custom firmware and resides in an Intel Multibus-based computer system. All programming of the system was done using IOForth (IO Inc., Tucson, AZ). The performance characteristics of the CID17-BAS camera system are summarized in Table 11.1.

TABLE 11.1. Slow-Scan CID and CCD Camera Specifications

Specification	CID17BAS (50 kHz pixel rate, -135 °C)	KAF1300L (500 kHz pixel rate, -100 °C)
Pixel size	23 μm(h) × 27 μm(v)	16 μm × 16 μm
Array dimensions	378(h) × 244(v)	1280(h) × 1024(v)
Photoactive area	8.7 mm × 6.6 mm	20.5 mm × 16.4 mm
Number of pixels	92,232	1,310,720
Read noise	100e$^-$ (100 NDRDs)	13e$^-$
Saturation charge	600,000e$^-$	200,000e$^-$
Peak QE (2)	47% (550 nm)	26% (700 nm)
QE at 225 nm	13%	12%
Linearity	0.7%	0.2%
Dark current	<0.5e$^-$/pixel/min	0.4e$^-$/pixel/min (-125 °C)
Blooming	Not observed at 12 Me$^-$/pixel/sec in 90 sec	Not observed at 5 Me$^-$/pixel/sec in 90 sec
Digitization	14 bits, operated w/ fixed gain	12 bits, 11 and 47e$^-$/ADU gain

11.4.2.2. KAF1300L Camera System

The charge-coupled device camera is based on an Eastman Kodak Company KAF1300L CCD. The CCD was introduced in January 1990 as a 1.3-million-pixel antiblooming full-frame imager. The camera system consists of a liquid-nitrogen-cooled camera head (model CH210), camera electronics module (model CE200), and a VME-bus camera controller (all from Photometrics, Ltd.). The camera controller board is used in a Sun 3e computer system (Sun Microsystems, Moutain View, CA). A camera controller board for a Multibus-based computer system (same model as above) has also been used in this work. The Multibus-based computer system was also configured by Photometrics (Model DIPS3000). The performance of the CCD camera system is summarized in Table 11.1 along with the CID performance data.

11.5. TRADITIONAL QUALITATIVE ELEMENTAL ANALYSIS WITH A MULTICHANNEL DETECTOR

An echelle polychromator with a solid-state two-dimensional detector array contains no moving parts and is therefore inherently stable. Wavelength calibration is easy to establish and maintain. A low-pressure mercury discharge is used as a reference source both to establish the initial calibration and to periodically verify it. The positions of the mercury lines are used to arrive at an empirical fit to both the prism and the grating dispersion. From these fits, wavelengths at other positions on the detector can be calculated. Verification of the calibration simply involves checking to see that the mercury lines fall on the correct pixels. Spectral lines for other elements are referenced via position on the detector relative to the mercury 253.65 nm line rather than by wavelength. This approach allows the location of a particular spectral line for an element by simply calculating an X and Y offset. Additionally, if the coordinates of the mercury line do change, the relative coordinates of the spectral lines for all of the other elements remain unaffected.

Qualitative analysis of a sample in the traditional sense can be carried out with varying degrees of sensitivity by controlling the exposure time of a single exposure.[21] A single image of the entire spectrometer focal plane is checked for the presence of emission at the locations corresponding to a particular element. With either the CID or CCD echelle spectrometer systems a relatively long exposure (1 minute) results in saturation of the intense lines from any element present at high concentration. Trace elements will show a number of spectral lines at intermediate intensities and ultratrace elements will show only a few of the most intense spectral lines at relatively low signal levels. Even if a spectral line is saturated, the qualitative analysis is unaffected since neither of the detectors blooms. Shorter exposure times (down to 0.2 second) are used to find only the major components of a sample.

11.6. TRADITIONAL QUANTITATIVE ELEMENTAL ANALYSIS

Quantitative elemental analysis in the traditional sense is performed using standardization, either internal or external. Spectral-line intensities are measured in terms of signal level per unit time, so that values derived from different exposure times can be directly compared. The differences in operation between the CID and CCD, specifically NDRO and pseudo-random access capability of the CID, lead to different means of implementation.

11.6.1. Random Access Integration with a CID Detector

The term, random-access integration (RAI), describes the combination of NDRO capability of the CID with the ability to pseudorandomly access individual pixels so that a number of spectral lines can be measured in a single exposure, each with a different integration time.[30] In this technique, small regions (3 pixels by 3 pixels) are centered on each of the spectral lines to be used for the analysis of a sample. NDROs are used to sample each of the 3- by 3-pixel regions in a continuous loop, once the exposure to the analytical source has begun. Photogenerated charge will accumulate at each of the regions at a different rate, depending on the concentration of the element in the sample and the strength of the emitter. NDROs are used to determine an optimum time for the calculation of the charge accumulation rate. At approximately 80% of the pixel saturation level, a number of NDROs (up to 100) are used to precisely measure the quantity of charge accumulated at the spectral line and in the background region adjacent to the spectral line (a 3- by 13-pixel region). The exact duration of the exposure is also recorded so that the signal level per unit time can be calculated. The exposure to the analytical source continues until all spectral lines have been recorded in a similar manner.

The concentration of an element in the sample is calculated based on the intensity per unit time measured at each spectral line of the element. Background correction is performed automatically by fitting a straight line to the baseline to either side of the spectral-line center. Blank subtraction is offered as an option and is performed after baseline correction of both the sample and the blank. Blank subtraction corrects for any trace contamination of the sample that may have occurred during sample work-up (reagents, laboratory environment). Baseline correction prior to blank subtraction corrects for long-term drift that may occur in the continuum background.

11.6.2. Variable-Integration Time Detection with an Antiblooming CCD

Quantitative analysis with an antiblooming CCD-equipped echelle polychromator is carried out in a manner similar to the method used with the CID echelle polychromator. The only exception is that, since NDRO capability does not exist, several separate exposures (two or three) of the complete focal plane must be made. A short exposure is used to measure the signal level per unit time for the intense spectral lines produced by elements present at high concentrations. A second, longer exposure is used to measure the intensity per unit time of the spectral lines from elements present at low concentrations. The term, variable-integration time detection (VIT), has been used to describe the extension of dynamic range through the use of multiple exposures, each of a different duration. The technique has been used with several types of detectors in the past, but to be successful with very wide dynamic range sources like atomic emission sources, blooming must not occur.

Three exposures can be used with the KAF1300L echelle polychromator to cover approximately 6 orders of magnitude in emission intensity. The camera provides 12 bits of intensity precision (4096 grey levels) and two programmable gain settings—a high gain setting of 11 electrons per count (*A*nalog to *D*igital converter *U*nit or ADU) and a low-gain setting of 64 electrons per ADU. The root-mean-square noise of the system is 15 electrons, so the noise is only observable at the high-gain setting. At the low-gain setting, a 0.2-second exposure covers charge-generation rates extending from 3200 electrons/sec $[(10 \text{ ADUs} \times 64 \text{ e}^-/\text{ADU})/0.2 \text{ sec}]$ to 1.28 million electrons/sec $[(4000 \text{ ADUs} \times 64 \text{ e}^-/\text{ADU})/0.2 \text{ sec}]$. A 5-second exposure, also at the low-gain setting, covers charge-generation rates from 128 electrons/sec to 51,200 electrons/sec. Finally, at the high-gain setting, a 1-minute exposure covers rates from 1.8 electrons/sec $[(10 \text{ ADUs} \times 11 \text{ e}^-/\text{ADU})/60 \text{ sec}]$ to 733 electrons/sec $[(4000 \text{ ADUs} \times 11 \text{ e}^-/\text{ADU})/60 \text{ sec}]$.

11.7. AUTOMATED QUALITATIVE ANALYSIS

Traditional qualitative elemental analysis by emission spectroscopy is usually done by looking for specific elements and using one or more spectral lines to confirm the presence or absence of an element. When the question is asked about only a few elements, the problem is easily handled with a slew-scanning monochromator or a fixed-slit polychromator, provided that channels are installed for the specific elements. When a large number of elements must be included in the survey, the problem becomes more difficult. Frequently the analyst is presented with a "good" and a "bad" material and is asked to find the difference. In these cases it is often desirable to perform a qualitative analysis for as many elements as possible.

The availability of the entire emission spectrum from a single exposure in an echelle polychromator equipped with a two-dimensional array detector makes qualitative surveys for a large number of elements relatively easy as described above. In the manual procedure the analyst simply calls up screen overlays to indicate the positions of the intense spectral lines for an element. One by one the analyst confirms the absence or presence of each element in the survey. The stability of the polychromator optical system leads to highly reproducible wavelength assignment to the screen overlays always line up.

11.7.1. Autoqual Algorithm

The qualitative analysis procedure is automated by replacing the decision making of the analyst with a computer algorithm. A flow chart of the algorithm is shown in Fig. 11.7. The process begins by acquiring an emission spectrum. For plasma emission spectroscopy, a blank subtracted emission spectrum is used so that emission features due to the argon plasma and solvent are not present. The corrected spectrum is produced by the digital subtraction of two images taken using the same exposure time, one of the sample solution and one of the blank, typically a 5% nitric acid solution. The algorithm uses a database containing information on up to ten spectral lines per element for the elements which can be determined by emission spectroscopy in the ultraviolet. The lines included in the database are the strongest emitters for each element (for a particular source type) and the position of each spectral line is stored in coordinates relative to the mercury calibration wavelength.

The algorithm checks for the presence of each element one at a time first by calculating the signal-to-noise ratio (SNR) at each of the spectral lines. A subarray of pixels centered on the spectral line location is used for both spectral line and background intensity measurements. The width of the subarray corresponds to the width of an echelle order and the length to approximately eight times the half-width of a spectral line. The resulting subarray is rectangular with a 4:1 aspect ratio. The two groups of pixels at either end of the subarray are used to fit the spectral background and to estimate the noise in the baseline. The spectral line intensity is calculated from the sum of the pixel intensities at the center of the subarray after the background contribution has been subtracted. The noise value used is the standard deviation of the mean of all of the background pixels.

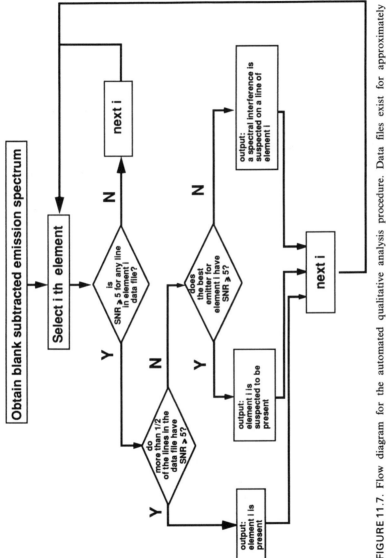

FIGURE 11.7. Flow diagram for the automated qualitative analysis procedure. Data files exist for approximately 60 elements, and these are searched sequentially to determine if the various criteria shown in the diagram are met. The "best emitter" is the spectral line which produces the highest signal level at the detector and incorporates detector sensitivity, spectrometer throughput, and source emissivity factors.

Once the SNR has been calculated for all of the spectral lines for an element, a check is made to determine if any of the SNRs exceed a predetermined value, typically 5. If none does then the algorithm moves on to the next element. The algorithm next checks to see if more than half of the lines have a SNR greater than 5. If more than half of the spectral lines exhibit a SNR greater than 5, then it is concluded that the element is definitely present. If fewer than half of the spectral lines have a SNR of 5 or more, then the algorithm checks to see if the SNR on the most intense emitter is over 5. If this is the case, then it can be concluded that the element is probably present, but just at such a low concentration that only the most intense lines have a SNR that is greater than 5. For these cases the algorithm reports that the presence of the element is suspected. If the most intense emitter does not have a SNR of 5 or more, then the line with the high SNR must be a less intense line. The only way that this can possibly occur is if there is a spectral interference. In these cases the algorithm reports that a spectral interference is suspected and outputs the SNR and the wavelength of the line.

One further refinement is required to ensure against false positives. A false positive can theoretically occur in one of two ways. If enough spectral interferences occur so that half of the spectral lines in an element database yield a SNR $\geqslant 5$, then the algorithm will report that the element is present. The likelihood of this occurring is kept low by making sure that each element database contains more than a few spectral lines and by using a polychromator with high resolution. A false positive can also be reported if the noise calculated for a given spectral line is anomalously low. Since the standard deviation is calculated from a relatively small number of pixel intensities, this outcome occurs frequently enough to be troublesome. The difficulty is avoided by tabulating noise values from all of the spectral lines found with a SNR $\geqslant 5$. Any spectral lines with a SNR $\geqslant 5$ and a noise value more than two sigma from the mean value are ignored.

An example of a typical automated qualitative analysis output for an element found to be present in the sample is shown in Table 11.2. The most intense emitter is placed first in the database for an element with the rest of the lines being in random order. The SNR value chosen as a cutoff value is shown in the output. Five has been found to be a conservative value which still provides high sensitivity. In Table 11.2 all of the spectral lines of chromium were detected at a SNR > 5, so the element was reported as being present in the sample. The other information in the table (and stored in the database) includes the line wavelengths and the coordinates of the spectral line relative to the mercury reference wavelength.

Table 11.3 shows an example of the output for an element not present in the sample. Even though a SNR $\geqslant 5$ is reported for the mercury line at 435.8 nm (line number 6 in the table), none of the other lines have a

TABLE 11.2. Automated Qualitative Analysis Report for
an Element Found to Be Present in a Sample[a]

Line	Element	Wavelength	Signal/noise	Signal	Bgnd. Dev.	ABS X	ABS Y
0	Cr	359.35	670.31	730.98	1.09	175	61
1	Cr	427.48	192.47	332.00	1.72	130	33
2	Cr	428.97	305.65	472.80	1.55	199	32
3	Cr	360.53	287.35	564.63	1.96	242	61
4	Cr	425.43	138.44	213.96	1.55	39	33
5	Cr	357.87	460.74	440.01	0.95	94	61

S/N Threshold = 5

Number of Lines = 6   Number Found = 6

[a] Synthetic solution containing 10 μg/mL each Cr, Ni, Al, Mg, and Sr analyzed by direct-current plasma emission.

SNR $\geqslant$ 5. This is not a case of the element being present at a low concentration since the most intense emitter, the 253.65 nm line (line 0 in the table), does not have a SNR $\geqslant$ 5. The example demonstrates how the automated qualitative analysis algorithm can be used to identify spectral interferences. By simply aspirating a single element solution, all other elements which have a spectral line exhibiting a SNR $\geqslant$ 5 must be spectral interferences. The routine also provides a measure of the severity of the interference, since the actual signal-to-noise rations are reported.

TABLE 11.3. Automated Qualitative Analysis Report
for an Element Not Found in a Sample[a]

Line	Element	Wavelength	Signal/noise	Signal	Bgnd. dev.	ABS X	ABS Y
0	Hg	253.65	−0.98	−4.9	5.08	214	168
1	Hg	404.66	−1.00	−4.2	4.22	151	36
2	Hg	365.00	0.05	0.24	5.30	160	54
3	Hg	313.20	0.22	0.66	3.05	134	90
4	Hg	312.57	0.83	3.18	3.81	97	89
5	Hg	296.73	1.21	5.82	4.80	151	104
6	Hg	435.83	8.50	38.1	4.49	126	26
7	Hg	366.33	−1.56	−4.56	2.93	232	52
8	Hg	365.48	0.38	1.44	3.80	186	52
9	Hg	546.00	−0.20	−0.86	4.24	272	2

S/N Threshold = 5

Number of lines = 10   Number found = 1

[a] Same solution as in Table 11.2. Note that the element is reported as not found even though one of the spectral lines gave a SNR greater than 5. Since the most intense emitter did not give a SNR greater than 5, the algorithm assumes a spectral interference has occurred.

The sensitivity of the automated qualitative analysis algorithm is demonstrated by analyzing a dilution series of a six-element standard solution. The series spans the concentration range from 5000 to 5 ng/g and contains the elements Ca, Cr, Mn, Fe, Ni, and Pb. Table 11.4 summarizes the results of the analysis. At the highest concentration all of the components are reported as being present, and also of equal importance; no false positives are reported. The greatest risk of a false positive occurs when the spectrum contains many intense spectral lines. As the solution becomes more dilute, the reported results change from present to suspected to be present to finally not found. The trend observed matches that predicted by the emission intensities of the spectral lines and the limits of detection for these elements with this instrument. For example, Pb, Fe, and Ni are reported as not found at a concentration of 5 ng/g, which is below the instrument detection limit for these elements. Also note the sensitivity which is still exhibited for the strong emitter Ca at the 5 ng/g level.

Even though a level of sensitivity close to the traditional quantitative analysis detection limit is achieved by the automated qualitative analysis algorithm, false positives are virtually nonexistent. The high level of discrimination is in large part due not only to precise background correction which is possible because of the simultaneous nature of the instrument, but also because spectral information from more than one emission line is being used.

11.7.2. Automated Qualitative Analysis Examples

Some potential applications for the automated qualitative analysis routine are illustrated with the following examples. Figure 11.8 shows the results of the qualitative analysis of three stream water samples. The samples were taken near a mining operation in southern Arizona. One sample was taken from upstream of the mining operation, one sample was taken of the runoff from the operation, and one sample was taken from downstream of the operation. By being able to simultaneously screen for

TABLE 11.4. Automated Qualitative Analysis Results
for a Dilution Series of a Six-Element Standard Solution

	Present	Suspect	Not found
5000 ppb	Ca, Cr, Mn, Fe, Ni, Pb	—	—
500 ppb	Ca, Cr, Mn, Fe, Ni	Pb	—
50 ppb	Ca, Cr, Mn	Fe, Ni, Pb	—
5 ppb	Ca	Cr, Mn	Pb, Fe, Ni

the presence of many elements, it is possible to quickly determine which elemental species serve as good chemical markers of the mining output. For example, the concentration of iron, cobalt, and copper in the mining effluent is very high, but no significant difference exists between the upstream and downstream water samples. This is due to the instability of iron, cobalt, and copper in the water column at the natural pH of the stream water. Manganese and zinc, however, can serve as markers because they are undetected in the upstream water sample, are present in the mining effluent, and are still detectable downstream. A decrease in concentration is observed, however, because of the dilution of the mining effluent.

Another strength of the automated qualitative analysis algorithm is that little or no prior knowledge about the sample is necessary. This is beneficial in applications such as the screening of unidentified laboratory waste solutions. The composition of the waste solutions must be determined so that the proper disposal method can be chosen. The results of an automated qualitative analysis of an unidentified aqueous waste solution from a university teaching laboratory is shown in Table 11.5. Once these results were reported to the laboratory that generated the waste, they were able to identify the processes which generated the waste. The likely source of the manganese was a still used to make triply distilled water and the tin and cadmium were used in several electrochemical experiments. Boron, calcium, magnesium, aluminum, and silicon were reported in the qualitative analysis and were verified by independent measurements to be present. The likely source of these elements is the glass container in which

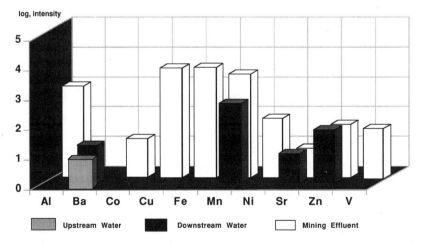

FIGURE 11.8. Automated qualitative results for a stream water analysis in the vicinity of a mining operation. Samples of the mining effluent, upstream water, and downstream water show which of the metals introduced by the effluent remain in solution.

TABLE 11.5. Automated Qualitative Laboratory
Waste Solution Analysis

Elements found to be present	Elements suspected to be present
Manganese	Magnesium
Boron	Cadmium
Calcium	Barium
Aluminium	Silicon
Tin	
Carbon	

the waste was collected. Two important points are illustrated by this example: the automated qualitative analysis was successful in analyzing an uncharacterized sample, and sensitivity comparable to conventional quantitative analysis is achieved as demonstrated by the detection of ions leached from the glass container.

Automated qualitative analysis is not limited to solution samples analyzed by plasma emission. By using a spark source for excitation, it is possible to use automated qualitative analysis for rapid screening of metal alloys. The results for samples of low-alloy steels and aluminum samples are shown in Tables 11.6 and 11.7. We note that there were no false positives reported even in the spectrally rich spark spectra. All elements were either positively identified, suspected to be present (trace alloying components whose concentration was so low as not to provide a $SNR \geqslant 5$ on 50% or more of the emission lines), or not detected. Those elements listed as not detected are present at a concentration near or below the limit of detection for the instrument or, in the case of sulfur, beryllium, and gallium, a database containing spectral line information for the element does not currently exist.

An additional refinement that could be added to the automated qualitative analysis routine to aid in the identification of elements would be to examine the relative emission intensity of the spectral lines for an element. Spectral lines would not only have to be found in a sample, but their relative intensities would have to be consistent given the sample type. Fortunately, this has not turned out to be necessary as the multiline approach appears to be sufficiently robust. If relative line intensities were considered, the algorithm would grow much more complex in order to avoid reporting false negatives. All spectral interferences would have to be corrected for in an iterative process before the ratios for an element could be correctly calculated. Additionally, variations in sample matrix composition significantly alter the source excitation temperature which, in turn, affects spectral-line-intensity ratios (particularly atom-line to ion-line ratios). Therefore, a robust qualitative analysis scheme which requires

TABLE 11.6. Automated Qualitative Steel Analysis

Standard[a]	Elements present	Elements detected	Elements suspected	Not detected
SS 406/1	Fe, C, Si, Mn, P, S, Cr, Mo, Ni, Co, Cu, V	Fe, C, Si, Mn, Cr, Mo, Ni, Cu, V	Co—0.006%	P—0.009% S—no database
SS 407/1	Fe, C, Si, Mn, P, S, Cr, Mo, Ni, Cu, V	Fe, C, Si, Mn, Cr, Mo, Ni, Cu, V	P—0.030%	S—no database
SS 408/1	Fe, C, Si, Mn, P, S, Cr, Mo, Ni, Cu, V	Fe, C, Si, Mn, Cr, Mo, Ni, Cu, V	P—0.037%	S—no database
SS 409/1	Fe, C, Si, Mn, P, S, Cr, Mo, Ni, Co, Cu, V	Fe, C, Si, Mn, Cr, Mo, Ni, Cu, V	Co—0.014% P—0.025%	S—no database
SS 410/1	Fe, C, Si, Mn, P, S, Cr, Mo, Ni, Cu, V	Fe, C, Si, Mn, Cr, Mo, Ni, Cu, V	P—0.072%	S—no database

[a] Bureau of Analyzed Samples, Ltd., Newham Hall, Middlesbrough, England.

emission-line-intensity rations to fit a predetermined pattern would have to take into consideration source excitation conditions as well. This is in contrast to ICP-mass spectrometry, where the iterative approach is quite successful because of relatively small variability in isotope ratios.[31]

Although qualitative screening is useful as an initial means of analysis, many applications require a better estimate of the concentration of each constituent. Simply using the SNR obtained from the automated qualitative analysis routine provides a crude estimate, but it is not highly reliable since the sample matrix can have such a pronounced effect on emission line intensities. For example, in the direct current plasma (the source used for the liquid sample analyses described above), the presence of easily ionized elements in the sample can lead to as much as a 20% enhancement of

TABLE 11.7. Automated Qualitative Aluminum Analysis

Standard[a]	Elements detected	Elements suspected	Elements missed
SS-1100-AN	Al, Mg, Si, Fe, Cu, Zn, Ti	Cr, Mn	Ni—0.003% Ga—no database
SS-2011-X	Al, Mg, Si, Fe, Cu, Cr, Zn	Mn, Cr, Ni	Ti—0.005%
SS-2618-F	Al, Mg, Si, Fe, Cu, Ni	Mn, Zn, Ti	Cr—0.001%
SS-2324-B	Al, Mg, Si, Fe, Cu, Mn	Cr, Zn, Ti	Ni—0.009% Be—no database Ga—no database

[a] Aluminum Company of America, Alcoa Center, PA.

atom-line emission intensities and a corresponding depression of the emission intensity from ion lines.[32] Changes in the nebulization conditions can also be caused by changes in the sample matrix, and these can lead to erroneous results. For example, a change in surface tension will alter the aerosol particle-size distribution which, in turn, will alter the mass per unit time of sample transported into the plasma. These problems associated with the composition of the sample matrix can at least partially be overcome by using an internal standard. An internal standard in this context is simply an element whose concentration in the sample is known or which can be spiked into the sample solution at a known concentration. Measuring analyte line intensities relative to the internal standard compensates for changes in sample aerosol introduction rate and sample excitation conditions. The internal standard can also serve as an ionization buffer making the various samples and standards under analysis more uniform in composition.

11.8. STANDARDLESS SEMIQUANTITATIVE ANALYSIS

The semiquantitative analysis algorithm was developed not only to give the qualitative composition of a sample, but also a better estimate of the concentration of each component. A semiquantitative type of analysis is advantageous when speed of analysis is more important that high accuracy. The semiquantitative algorithm is also used as the first step in a matrix-dependent spectral-line-selection routine for true quantitative analysis as described in a subsequent section. An internal standard element is used and is ideally chosen to match the spectroscopic, physical, and chemical

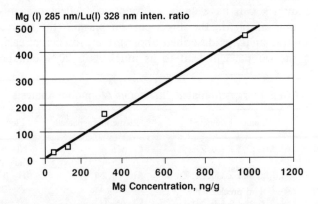

FIGURE 11.9. Emission-ratio working curve for magnesium using lutetium as the internal standard. The lutetium emission intensity at 328 nm from a 100 μg/ml spike is divided into the magnesium emission intensity at 285.213 nm. The figure shows linear behavior for Mg concentrations from 30 ng/ml to 1 μg/ml.

behavior of the analyte element or elements. Using an echelle polychromator with a two-dimensional multichannel array, internal standard and analyte lines can be monitored simultaneously so that there is no increase in analysis time and source drift is less of a problem. The ratio of the analyte response to that of the internal standard compensates for fluctuations in sample transport parameters and ideally corrects for nonspectral interferences that affect the analyte and internal standard similarly. Factors used in choosing an appropriate internal standard are discussed by Ahrens and Taylor[33] and also by Slavin.[34]

In solid sample analysis, the matrix material serves as the internal standard. In liquid samples, an element is typically added to both the sample and the standard solutions which are used initially to determine the correspondence between emission-intensity ratio and concentration. In liquid solutions, the internal standard can also serve as a spectroscopic buffer—a material used to partially swamp out the effects of variations in the concentration of easily ionized elements between samples. Once the equations for the emission-ratio working curves are obtained, semiquantitative information about the sample can be obtained over an extended period without the need to recalibrate. Figure 11.9 is an example of an emission-ratio working curve for magnesium with lutetium as the internal standard using a DCP for excitation. Figure 11.10 shows the results of the semiquantitative evaluation of the magnesium concentration in a solution determined daily over a 15-day period. The solid line represents the magnesium concentration determined with 10 replicate analyses using

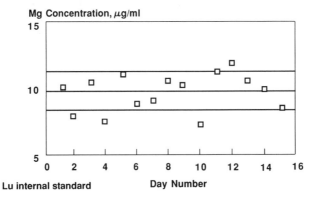

FIGURE 11.10. Long-term precision and accuracy study for semiquantitative determinations of magnesium concentration using lutetium as an internal standard in the direct current plasma. The test solution was prepared with a "true" concentration of $10\,\mu g/ml$ Mg ($100\,\mu g/ml$ Lu internal standard) as verified by 10 replicate analyses using separately prepared standard solutions. The study was carried out over 15 days using 3 replicate analyses per day without standardizing the instrument.

standards in a traditional quantitative analysis. The points are the values which were obtained over the following 15 days. Care was exercised to keep the instrument and analysis conditions as uniform as possible over the test period; however, no calibration was performed for any of the analyses. As the figure shows, reproducible semiquantitative results were obtained with the results never varying by more than $\pm 20\%$ over the 15-day period.

Table 11.8 shows a comparison of the reported (Thorn Smith, Royal Oak, MI) concentration values for alloying components in steels versus those obtained with the semiquantitative algorithm using the DCP for excitation. Iron emission at 259.940 nm was used as the internal standard. Although the accuracy achieved may not be acceptable in some applications, the advantage of not having to prepare calibration standards more than offsets the small reduction in accuracy for other applications. Examples of applications for which the accuracy of semiquantitative analysis might be sufficient include quality assurance where the technique could be used to ensure that the concentration of certain elements does not exceed some specified value. Another potential application might be as a screening technique for rapid and reliable identification or sorting of materials. Other frequently encountered potential applications include problem solving, where differences in composition are being sought, and in screening raw materials for new manufacturing processes or from new vendors for potential contaminations.

A question that is raised when using the semiquantitative analysis algorithm, particularly when analyzing materials such as steels which produce an emission spectrum with a very large number of intense spectral lines, is: do inaccuracies arise simply as random variations or do uncorrected

TABLE 11.8. Semiquantitative Analysis of Steels

Standard[a]	% Mo	% Mn	% Si	% Ni	% Cr
36 present	0.18	0.96	0.24	0.03	0.95
found	0.21	0.89	0.27	0.03	1.05
38 present	0.0	1.22	0.01	0.0	0.0
found	0.0	1.18	0.006	0.0	0.0
40 present	0.23	0.68	0.25	1.81	0.70
found	0.19	0.73	0.22	1.77	0.67
42 present	0.19	0.83	0.27	0.14	1.07
found	0.23	0.88	0.25	0.12	1.21
47 present	0.0	0.05	0.74	10.41	18.35
found	0.0	0.02	0.69	9.67	17.63
53 present	0.01	0.34	0.02	0.01	0.01
found	0.003	0.28	0.05	0.02	0.006

[a] Thorn Smith, Royal Oak, Michigan.

spectral interferences cause true biases? The solution to this dilemma, and the next step in the evolution of the intelligent atomic emission spectrometer which performs completely automated analysis, is the development of an algorithm for choosing the best spectral lines for a particular analysis and for accurately correcting for any remaining spectral interferences. Thus far, an algorithm has been described for qualitatively identifying the composition of a sample and then roughly determining the concentration of each component, so all of the information is available for the next step.

11.9. MATRIX-DEPENDENT SPECTRAL-LINE SELECTION

The proposed approach for matrix-dependent spectral line selection is that put forth by Boumans et al.[35,36] A "true detection limit" is defined as the sum of the conventionally determined detection limit and a selectivity term which accounts for additional signal (noise) introduced by the presence of interfering spectral lines. Following the notation used by Boumans, the true detection limit is given by

$$C_{L,\text{true}} = \tfrac{2}{5} \sum \{[S_{I,j}(\lambda_a)]/S_A\} \, C_{I,j} + C_{L,\text{conv}} \tag{11.1}$$

where $C_{L,\text{true}}$ is the true limit of detection, $C_{L,\text{conv}}$ the conventional limit of detection, $C_{I,j}$ the concentration of interferant j, $S_{I,j}(\lambda_a)$ the sensitivity of the interfering line from element j at the analytical wavelength λ_a, and S_A is the sensitivity of the analyte line.

The best spectral line for a determination is the one which gives the lowest $C_{L,\text{true}}$. The information required to determine which of the prominent lines will give the best detection limit in a particular sample is the sensitivity (S_A) of the analyte lines and the sensitivity of all the potential interferants, $S_{I,j}(\lambda_a)$, as well as the concentration of the interferants. As a matter of convenience, the sensitivity ratios are calculated and stored rather than the individual sensitivities. Boumans calls these quotients Q values:

$$Q_{I,j}(\lambda_a) = S_{I,j}(\lambda_a)/S_A \tag{11.1}$$

The Q ratios are more valuable because they are independent of spectrometer throughput and detection efficiency; however, source excitation conditions and spectrometer resolution do have an impact. The Q values are easily determined using a multichannel-detector-equipped polychromator. Simply by running a series of pure standards (100 μg/g, for example) and monitoring the intensities at all of the spectral lines for the other elements, the Q values can be determined readily.

TABLE 11.9. Sensitivity and Q Values for Ni in a Steel Matrix[a]

Wavelength	Ni sensitivity ADU/ppm	Q—iron	Q—molybdenum	Q—manganese	Q—chromium	Q—vanadium
349.30 nm	5.1	38	1262*	18	55	243*
341.47 nm	9.3	4918	6728	62	101*	14500
352.45 nm	5.3	1492*	5011*	121	1869*	37
339.30 nm	4.3	1952*	1113*	211	10013	32
344.63 nm	4.2	318	7909*	0	8211*	342
347.25 nm	4.1	3332	285	589	56367	615*

[a] Analysis by direct current plasma emission with the CID17-BAS echelle spectrometer.
Values marked with an asterisk denote spectral interference on the wing of the analytical line. Background correction is determined using a straight-line fit to the background on only one side of the analyte line.

Sensitivity and Q values have been tabulated in our laboratories for the major alloying components in steel (iron, manganese, chromium, nickel, molybdenum, and vanadium) and also for the lanthanide elements. An example of the tabulated values for alloying components of steel is shown in Table 11.9, which shows the sensitivity and Q values for nickel in a steel matrix. The sensitivity is given in units of ADUs per $\mu g/g$ and the Q values are the sensitivity ratio $\times 10^6$. The values marked with an asterisk were obtained using a straight-line fit to the background data from only one side of the analyte spectral line (the normal procedure is to use both sides). This was necessary because interferences occurred in the spectral line wing and higher-order background fits have yet to be implemented on the system.

An example of the application of the matrix-dependent spectral-line selection is in the analysis of a monazite sand. Monazite sands are rich in rare earth elements as shown in Table 11.10, which lists the typical

TABLE 11.10. Monazite Sand Typical Composition[a]

Ce	44.0	Yb	0.25
La	18.1	Tb	0.25
Nd	13.1	Er	0.19
Pr	4.8	Ho	<0.1
Y	3.3	Lu	<0.1
Sm	3.0	Tm	<0.1
Gd	2.2	Th	6.2
Dy	0.9	U	0.6
Eu	<0.5		

[a] Percent as oxides, Prof. Q. Fernando, University of Arizona, Department of Chemistry, personal communication.

composition. Lanthanides are spectrally rich and consideration of spectral interferences is essential. For example, the large amount of cerium present creates a problem for the determination of other rare earth elements. Typically, in the analysis of neodymium and lanthanum, the Nd line at 430.357 nm and the La line at 433.373 nm are the most sensitive lines and are commonly used for quantitative determinations. However, with the high cerium content of the monazite (44 % Ce_2O_3), these lines are interfered with. Figures 11.11a and b show working curves for neodymium and

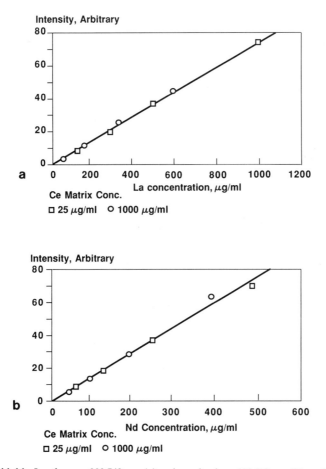

FIGURE 11.11. Lanthanum 333.749 nm (a) and neodymium 415.608 nm (b) working curves determined in the presence of cerium. Solid circles were obtained at a 25 μg/ml concomitant element concentration and open circles were obtained at a 1000 μg/ml concomitant element concentration. The good fit to a single straight line indicates freedom from spectral interference.

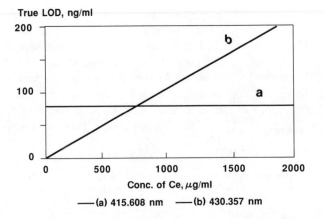

FIGURE 11.12. The limit of detection (LOD) for neodymium at two spectral lines as a function of cerium concentration in the sample matrix. The 430.357 nm spectral line is more sensitive when the cerium concentration in the matrix is low; however, the 415.608 nm line is more sensitive at high cerium concentrations because it suffers less from spectral interference.

lanthanum at alternate lines, 415.608 nm and 333.749 nm, respectively. The solid points represent data obtained in a matrix low in cerium (25 µg/g) and the open points represent data obtained in a matrix of 1000 µg/g Ce. All of the points fall on the same line, demonstrating that these lines are essentially interference free. However, automatically using the interference-free spectral line for an analysis, that is, selecting analyte lines based on Q values alone and ignoring the actual concentration of the interferant, ignores the inherently higher sensitivity of the line being rejected. Thus, the true detection limit calculation includes the interferant concentration term $(C_{I,j})$. The determination of Nd in a cerium matrix illustrates the impact of interferant concentration on optimum analyte line selection. At high Ce concentrations, the line at 415.6 nm is the best line; however, as the Ce concentration decreases, the line at 430.4 nm, which has a conventional limit of detection 10 times lower than the line at 415.6 nm, becomes the best line for analysis. The situation is shown graphically in Fig. 11.12. The crossover point occurs at a Ce concentration of approximately 700 µg/g and is dictated primarily by the resolution of the instrument.

11.10. MULTICHANNEL ATOMIC EMISSION EXPERT SYSTEMS IN THE NEAR FUTURE

In the very near future, the combination of automated qualitative analysis, semiquantitative analysis, and automatic spectral line selection will be routinely used for analysis. It is easy to envision an instrument which will use a single continuous process to first qualitatively identify all of the components present in a sample, then make estimates of the concentrations of each component, and, based on the estimates, choose the very best spectral lines for the determination of the concentrations of the analytes of interest. An analytical report for the analysis of a series of similar samples might contain information on differences in matrix composition which resulted in different spectral lines being used in addition to the concentrations of the analytes. Since the "true" detection limit for an element depends on the matrix composition of the sample, concentrations might be reported relative to the detection limit in the sample, rather than relative to detection limits in pure water as is sometimes done in ICP-AES.

Information relating to variation in matrix composition that is dis-covered by the "intelligent" atomic emission spectrometer has an impact on the accuracy of qualitative determinations by plasma emission as well as being of potential importance in problem solving or process control applications. Unanticipated variations in matrix composition may signal a problem occurring in the sample preparation steps. Assumptions regarding the degree to which the matrix composition of standards matches that of the samples may be invalidated. In these examples, the sample preparation procedure or the matrix composition of the standards may need to be altered to give the best possible analytical results. The time-consuming and expensive procedure of retesting samples which produce unexpected results could be avoided if there were a high degree of confidence in the original analysis.

A whole range of possibilities exist for improving the accuracy of analysis (removing sample specific biases) through making use of spectral features which have diagnostic value. The emission lines for the elements normally determined by atomic emission can be used to determine sample matrix composition, but other spectral features can be used to monitor sample introduction conditions and analyte excitation conditions. For example, in argon plasmas argon-atom to argon-ion intensity ratios indicate excitation temperature, hydrogen or OH emission correlate with aqueous sample introduction rates, carbon and hydrogen emission correlate with the concentration of organic species in the sample, etc. The use that is made of this additional information, if provided, is currently the subject of active research. At the least, abnormal conditions would be signaled so that corrective actions could be taken by an operator. More sophisticated approaches would apply correction factors when the correspondence is known. An alternative approach would strive to alter analysis conditions (powers, flow rates, etc.) so as to maintain the analytical source at constant conditions (excitation, sample introduction).

A revolutionary step has been taken in the level of sophistication of the hardware for elemental analysis by optical emission. The information available about a sample and the conditions under which it was analyzed has been increased by several orders of magnitude. The next step to be taken is to make rational use of all the additional data.

ACKNOWLEDGMENTS

The authors would like to acknowledge Dr. David Ekimoff and Mr. John Tritten for many insightful discussions and helpful comments, and Ms. Joanne Weber, Ms. Barbara Coccia, and Mr. Thompson Davis for the careful preparation of the final manuscript.

REFERENCES

1. P. W. J. M. Boumans, ed., *Inductively Coupled Plasma Emission Spectroscopy—Part 1 Methodology, Instrumentation, and Performance*, Wiley, New York (1987).
2. P. W. J. M. Boumans, ed., *Inductively Coupled Plasma Emission Spectroscopy—Part 2 Applications and Fundamentals*, Wiley, New York (1987).
3. P. W. J. M. Boumans and J. J. A. M. Vrakking, "Spectral Interferences in Inductively Coupled Plasma Emission Spectrometry I. A Theoretical and Experimental Study of the Effect of Spectral Bandwidth on Selectivity, Limits of Determination, Limits of Detection and Detection Power," *Spectrochim. Acta* **40B**, 1085–1105 (1985).
4. P. W. J. M. Boumans and J. J. A. M. Vrakking, "Spectral Interferences in Inductively Coupled Plasma Atomic Emission Spectroscopy II. An Experimental Study of the Effect of Spectral Bandwidth on the Inaccuracy in Net Signals Originating from Wavelength Positioning Errors in a Slew-Scan Spectrometer," *Spectrochim. Acta* **40B**, 1107–1125 (1985).
5. J. W. Olesik, L. J. Smith, and E. J. Williamsen, "Signal Fluctuations Due to Individual Droplets in Inductively Coupled Plasma Atomic Emission Spectrometry," *Anal. Chem.* **61**, 2002–2008 (1989).
6. J. W. Olesik and E. J. Williamsen, "Simultaneous Detection of One-Dimensional Laser-Induced Fluorescence or Laser Light Scattering Images in Plasmas," *Appl. Spectrosc.* **43**, 933–940 (1989).
7. J. W. Olesik and E. J. Williamsen, "Easily and Noneasily Ionizable Element Matrix Effects in Inductively Coupled Plasma Optical Spectrometry," *Appl. Spectrosc.* **43**, 1223–1232 (1989).
8. J. W. Olesik and A. W. Moore, Jr., "Influence of Small Amounts of Organic Solvents in Aqueous Samples on Argon Inductively Coupled Plasma Spectrometry," *Anal. Chem.* **62**, 840–845 (1990).
9. R. S. Pomeroy, J. D. Kolczynski, J. V. Sweedler, and M. B. Denton, "Analysis of Microgram Amounts of Particulate Material by Simultaneous Multiwavelength AES," *Mikrochim. Acta III*, 347–353 (1989).
10. L. A. Fernando, "Figures of Merit for an ICP-Echelle Spectrometer System," *Spectrochim. Acta* **37B**, 859–868 (1982).
11. R. B. Myers, "A New Concept in ICP-AES: A Multichannel Polyscanning Spectroanalyzer," *Spectroscopy* **4**, 16–24 (1989).

12. V. Karanassios and G. Horlick, "Spectral Characteristics of a New Spectrometer Design for Atomic Emission Spectroscopy," *Appl. Spectrosc.* **40**, 813–821 (1986).
13. G. M. Levy, A. Quaglia, R. E. Lazure, and S. W. McGeorge, "A Photodiode Array Based Spectrometer for Inductively Coupled Plasma-Atomic Emission Spectrometry," *Spectrochim. Acta* **42B**, 341–351 (1987).
14. Y. Talmi, "Applicability of TV-Type Multichannel Detectors to Spectroscopy," *Anal. Chem.* **47**, 658A–670A (1975).
15. D. F. Barbe and S. B. Campana, "Imaging Arrays Using the Charge-Couped Concept," in: *Advances in Image Pick-up and Display* (B. Kazan, ed.), Vol. 3, Chapter 3, Academic Press, New York (1977).
16. P. K. Weimer and A. D. Cope, "Image Sensors for Television and Related Applications," in: *Advances in Image Pick-up and Display* (B. Kazan, ed.), Vol. 6, Chapter 3, Academic Press, New York (1983).
17. A. J. Steckl, "Charge-Coupled Devices," in: *The Infrared Handbook* (W. L. Wolfe and G. J. Zissis, eds.), Chapter 12, Office of Naval Research, Dept. of the Navy, Washington, DC (1978).
18. E. L. Dereniak and D. G. Crowe, *Optical Radiation Detectors*, Wiley, New York (1984).
19. G. R. Sims and M. B. Denton, "Characterization of a Charge-Injection Device Camera System as a Multichannel Spectroscopic Detector," *Opt. Eng.* **26**, 1008–1019 (1987).
20. G. R. Sims and M. B. Denton, "Spatial Pixel Crosstalk in a Charge-Injection Device," *Opt. Eng.* **26**, 999–1007 (1987).
21. R. B. Bilhorn and M. B. Denton, "Elemental Analysis with a Plasma Emission Echelle Spectrometer Employing a Charge Injection Device (CID) Detector," *Appl. Spectrosc.* **43**, 1–11 (1989).
22. I. Fujii, "Semiconductor Charge-Coupled Device With an Increased Surface State," U.S. Patent 4, 742, 381 (1988).
23. H. Yamazaki, "Anit-Blooming Charge Transfer Device," U.S. Patent 4, 328, 432 (1982).
24. P. A. Levine, "Blooming Control for Charge Coupled Imager," U.S. Patent 3, 931, 465 (1976).
25. J. R. Janisick, Interlaboratory memos, April 23 to May 10, 1990, Jet Propulsion Laboratory, Pasadena, CA.
26. P. N. Keliher and C. C. Wohlers, "Echelle Grating Spectrometers in Analytical Spectrometry," *Anal. Chem.* **48**, 333A–340A (1976).
27. D. J. Schroeder, "Design Considerations for Astronomical Echelle Spectrographs," *Publ. Astron. Soc. Pacific* **82**, 1253–1275 (1970).
28. W. Cash, "High Efficiency Spectrographs for the EUV and Soft X-Rays," *IEEE Trans. Nucl. Sci.* **NS-30**, 1535–1538 (1983).
29. R. B. Bilhorn, *Analytical Spectroscopic Capabilities of Optical Imaging Charge Transfer Devices*, Ph. D. Dissertation, University of Arizona, Tucson, AZ (1987).
30. R. B. Bilhorn and M. B. Denton, "Wide Dynamic Range Detection with a Charge Injection Device (CID) for Quantitative Plasma Emission Spectroscopy," *Appl. Spectrosc.* **44**, 1538–1546 (1990).
31. D. Ekimoff, A. M. Van Norstrand, and D. A. Mowers, "Semiquantitative Survey Capabilities of Inductively Coupled Plasma Mass Spectrometry," *Appl. Spectrosc.* **43**, 1252–1257 (1989).
32. M. H. Miller, D. Eastwood, and M. S. Hendrick, "Excitation of Analytes and Enhancement of Emission Intensities in a D.C. Plasma Jet: A Critical Review Leading to Proposed Mechanistic Models," *Spectrochim. Acta* **39B**, 13–56 (1984).
33. L. H. Ahrens and S. R. Taylor, *Spectrochemical Analysis*, 2nd ed., Addison-Wesley, Reading, MA (1950).
34. M. Slavin, *Emission Spectrochemical Analysis*, Wiley Interscience, New York (1971).

35. P. W. J. M. Boumans and J. J. A. M. Vrakking, "Detection Limit Including Selectivity as a Criterion for Line Selection in Trace Analysis Using Inductively Coupled Plasma-Atomic Emission Spectrometry (ICP-AES)—A Tutorial Treatment of a Fundamental Problem of AES," *Spectrochim. Acta* **42B**, 819–840 (1987).
36. P. W. J. M. Boumans, J. A. Tielrooy, and F. J. M. J. Maessen, "Mutual Spectral Interferences of Rare Earth Elements in Inductively Coupled Plasma Atomic Emission Spectrometry—I. Rational Line Selection and Correction Procedure," *Spectrochim. Acta* **43B**, 173–199 (1988).

Index